ORIGINATIONS OF LIFE

from

VOLCANOES AND PETROLEUM

A Scientific Theory Opposed to Evolution

by

Donald E. Tyler

1092 Southwest Second Avenue
Ontario, Oregon 97914

Cover by I. Lambert
Printed by Ryan Gwinner Press, Portland, Oregon

dedicated to my mother

IVA HESS TYLER

without whose love, help, and encouragement
this contribution could not have been made

CONTENTS

I. INTRODUCTION

The primary subject of this book is the beginning of life. In the new theory to be presented, petroleum was the source of important chemicals essential to living organisms. In order to understand how petroleum originated, it is essential to understand the cause of volcanoes. It is necessary to understand the relationships between volcanoes and petroleum to understand how petroleum has been present in certain volcanic eruptions. The cause of volcanoes and how petroleum is formed presented here are new simple concepts.

The new theory of multiple originations of life is also a simple concept. It is a far more simple process by scientific standards than the theory of evolution. As much as possible, observations that anyone can verify will be used for proof and to aid in understanding the theory.

Presently accepted concepts of volcanoes, petroleum, and origins of life are complicated, confused, and characterized by inconsistencies. Those inconsistencies will be presented only to the extent believed necessary to show that presently accepted concepts are wrong and that the original theory presented herein is probably true.

This theory deals with and accounts for the beginnings of all the various and complex forms of life. The theory in vogue for more than a century is that life began as a simple organism from which all plants and animals developed. It will be shown to be inconsistent with the facts. It is inconsistent with scientific observations and experiments, and it is inconsistent with observations of plants and animals that anyone can make. This original scientific theory of multiple creations of living organisms will be shown to satisfactorily explain relevant observations. It will be shown that the new theory is probably true.

This is a scientific theory of volcanoes, petroleum, and life. As a scientific theory, it is a proposed solution, reason, or cause for facts observed. It simplifies, explains, accounts for, and encompasses all pertinent observations. It can serve as a working tool for systematic, reasoned investigation and systematic accumulation of knowledge. It is a reasoned principle which is less confusing, more easily remembered, and more useful than are disconnected observations. The purpose of the theory is to aid in understanding underlying principles of what has occurred, what presently occurs, and what may be expected to occur in the future. It will aid in understanding causal relationships between the environment of the Earth and the nature, character, and existence of life.

It has been claimed that Darwinism has been established as firmly as Newton's "laws." No scientific theory is perfect or immune to correction. Newton's "law" of gravity has been thoroughly refuted in my *New and Simple Theory of Gravity.* Newton lacked information of atomic structure and various observations including the expanding nature of the universe and mass limit, the limitation of the size that an accumulation of matter may attain. As a result, instead of a postulated force that encompasses all observations related to gravitation, other forces including a so-called nuclear force have been postulated and the bizarre "big-bang" theory has been added. Universal attraction by a gravitational force as proposed by Newton would result in all matter being in one atom, in one big mass, or moving toward such a status. As a result of Newton's not understanding atomic structure a confused picture has resulted with multiple theories to explain inconsistencies when a simple theory explaining gravitation based upon atomic structure and fundamental principles of physics encompasses all relevant observations.

In this presentation the postulate of evolution is challenged as well as the various proposed forces causing it including Darwin's hypothesis of natural selection of the fittest. Just as Newton lacked information of atomic structure, Darwin lacked information now available of cellular structure, cellular physiology, and genetic processes. A theory that is *truly a scientific theory* is tentative subject to further observations and predictions with following experience such as fulfillment or non-fulfillment of those predictions. It is submitted that Darwin and other evolutionists have errored and it is time to reevaluate and correct their theories.

The basic theory accepted by the scientific community of relationships between living organisms is the theory of evolution which embraces the postulate that the simplest form of life originated by chance. Its basic tenet is that more complex organisms resulted from changes in succeeding progeny of an original simple organism. Various sources for life's substances and how life started have been postulated. Various forces or causes to explain evolution have also been postulated. The issues are not decided.

Until approximately sixty years ago, it was generally accepted by scientists that life originated by chance combination of chemicals in the ocean. The idea was subject to considerable criticism because the mathematical probability or chance of the necessary complex chemical reactions to occur in such a weak, dilute solution is essentially zero. That chance or probability was recognized to be so low that the theory of evolution was necessary to account for the variety of living organisms by mechanistic means. To postulate that the chemical combinations essential for life could occur more than once in the oceans would be completely unbelievable. It would be like getting the same number a million times in succession on an unwired roulette table and then repeating the sequence again and again.

A.I. Oparin, a Russian, proposed a new and different theory of the origin of life in 1924. He had knowledge of experiments in petroleum formation by D. Mendeleyev which had shown that steam acting on carbides, metal combinations with carbon, produced acetylene. Acetylene is a hydrocarbon, a chemical combination of hydrogen and carbon. It is a compound that readily reacts to produce larger hydrocarbons such as sugars.

Oparin postulated that the Earth condensed from an extremely hot mass of gas. As it contracted, carbides flowed to the surface. There they came in contact with an atmosphere of hot steam. Hydrocarbons resulted as in Mendelyev's experiments. Nitrogen carbide (C_2N_2) observed in comet tails was then postulated to have combined with hydrocarbons from earth carbides to form amino acids. Those molecules floated, attracted each other to become larger molecules, and then they gathered in loose masses as colloid gels. Those gels formed small globs, individual small structures. That set them apart from the ocean. They absorbed more chemicals, grew larger, divided, and eventually became living cells. From those cells all living organisms evolved.

By 1961, Oparin had modified his theory to encompass current ideas. He proposed that as the Earth formed, it released methane and ammonia as gases. Those gases reacted with water in the atmosphere to form amino acids. The amino acids precipitated with rain, and then concentrated in the oceans where the chemical reactions necessary to form life occurred. He retained the idea of colloid globs or "coacervates" as the basis of cell structure.

There is no evidence in sediments of the primitive colloidal gels Oparin postulated. His first theory was proposed at a time when proteins were considered to be the essence of life. The nucleus of the cell, deoxyribonucleic acids, and genetics were not understood. His postulate of colloidal gel "coacervates" as the early cells of life seems primitive and fanciful, but the idea of such a structure to form a surface template upon which to build a real membrane may have merit. Further, his concept of relationships of carbon in the earth, petroleum, and life has a general similarity to the new theory I am proposing.

In 1929, J.B.S. Haldane postulated that the primitive atmosphere was mostly carbon dioxide, CO_2, and ammonia, NH_3. Ultraviolet light, in his theory, caused those molecules to react to form sugars and materials for proteins. The complex chemicals formed in a "hot dilute soup." Out of that soup large molecules formed and became enclosed in an oily film. From that structure the primitive cell of the first living organism formed. All living organisms evolved, according to him, from that one ancestor.

In 1967, J.D. Bernal proposed another theory based upon a postulated similar atmosphere. He postulated that the primitive atmosphere was composed of water, H_2O, methane, CH_4, and ammonia, NH_3. Experiments by Miller and Urey in 1953 and by others had shown that electric sparks simulating lightning in a gaseous mixture of those compounds could result in amino acids, sugars, and other compounds. Bernal postulated that such occurred in an atmosphere that contained no oxygen and then precipitated with rain into the oceans. To get the concentrations essential to form larger more complex chemicals he postulated adsorption of the early hydrocarbons, including amino acids, onto particles of clay. Then small pools containing those clumps of amino acids and other hydrocarbons evaporated. Concentrations of the hydrocarbons took place in mud, not open water. Those small hydrocarbons polymerized, that is, joined together, to form the more complex chemicals of life including nucleic acids, proteins, and sugars. It was postulated that through evolution cells were organized from those chemicals.

Miller and Orgel, in 1974, contributed another modification of the foregoing theories. When the Earth formed, heat from its interior caused "outgassing" through volcanoes. As a result the atmosphere was composed mainly of methane, ammonia, and water. Lightning as the major source of energy caused those molecules in the atmosphere to react with each other to produce amino acids and other hydrocarbons. They recognized that polymerizations were not likely to have occurred in the bulk of the oceans, so evaporation of lakes and tidepools were considered to have been important. The inconsistency was noted that some hydrocarbons were more volatile than water and would evaporate first, frustrating the concentration. Adsorption onto clays, absorption into colloidal particles, freezing, and primitive lipid (fat-like) membranes were considered to have been possible concentrating mechanisms.

The foregoing theories have problems. They postulate an atmosphere with water as a major constituent. It is generally agreed that ultraviolet light acts upon water in the upper atmosphere to produce free oxygen. It has been impossible to produce the necessary chemicals for life that have been produced with oxygen present. Other problems involve the dilution of ammonia and methane with nitrogen in the atmosphere and lack of high temperatures necessary to obtain certain essential chemical reactions. Critical and devastating, it has been impossible for the chemist-proponents of those theories to produce certain essential chemicals, to be presented in more detail, under conditions simulating their postulated primordial conditions.

The starting source for the chemicals of life in this theory is the substance of the Earth. It is postulated that heat in the depths of the Earth caused the material of magma to be vaporized and the resulting gases formed petroleum. It is a proven possibility, and it will be shown how it probably occurred.

Complex chemical compounds composed mainly of carbon and hydrogen, with many other elements in smaller amounts, exist in petroleum which could form the basic chemicals associated with life with relatively minor changes. The chemicals needed are concentrated in petroleum to a degree not duplicated or possible elsewhere in nature. That concentration is necessary for the chemical reactions to occur which are essential to life. Furthermore, heat, pressure, catalytic clays, and radioactivity occur with petroleum formation as proposed here. Those are conditions and factors that cause chemical reactions, including polymerizations, to occur with efficiency.

Petroleum could provide necessary complex chemical compounds required to produce life, and it could also provide nutrients to sustain it. Not only can bacteria and yeast grow on petroleum, they can convert it to protein efficiently on a commercial basis. Yields of protein weight as much as 50% of the weight of the petroleum used have been reported.

The real or absolute essence of life is unknown. Complex chemicals are invariably associated with life and are probably essential to life, at least as to life that we recognize. Some such chemical compounds have been shown to carry the code or pattern for the entire organism. That code includes virtually everything about the organism, including its entire anatomy (structure), physiology (functioning of body parts), maturation and reproduction. If life is to be understood, and particularly its origin, it is logical that an understanding of its chemicals is a reasonable starting place.

This new theory is based upon the postulate that petroleum is the intermediate source of chemical compounds essential for life, the primary source being the substance of the Earth. Of course, if the origin of petroleum is *entirely* from formerly living organisms, it could not serve the function of supplying the ingredients for the genesis of the earliest living organisms.

Evidence and reasoning will be presented that will establish that petroleum is probably inorganic and abiotic (not biologic) in origin. The theory is that gases arose from within the hot mass of the planet. It is proposed that some of those gases have escaped through volcanoes while some of the gases have been and continue to be trapped in sands and gravels. Some of the trapped gaseous molecules are postulated to have joined together to form the complex chemicals of petroleum.

It is postulated that hot, freshly produced petroleum and gases released through common natural seepages and certain volcanoes provided some of the conditions necessary to form the chemicals of life. Since it is most logical that the simplest were formed before the complex, the earliest forms of life are reasonably postulated to have been the simplest forms of life. With petroleum and its gases providing an abundant source of materials through natural common seepages and some volcanoes, where one such organism could have been formed, it is postulated that in order to achieve the required chemical reactions the conditions had to be such that many such simple organisms would have been produced. Some of those simplest organisms grouped together, possibly held together by a surrounding film of oil and other chemicals. Single complex cells resulted. Complex chemical reactions usually produce a variety of substances, which probably occurred. Further, minor chemical differences in the various petroleums, minor chemical differences in clays and rocks, and differences in envirooment such as temperature and humidity caused differences in chemical reactions, differences in resulting chemical compounds, and finally differences in the resulting single-celled organisms.

It is postulated that the early single cells formed included germ cells, including cells that divide asexually, cells with both male and female parts necessary for reproduction within the same cell, and male and female germ cells such as ova and sperm. Those single-celled germ cells produced complex organisms. Mixtures of chemical compounds produced in chemical reactions and minor changes in chemicals, heat, pressure, catalysts, or radioactivity present caused a variety of germ cells to be produced. The result of fertilizations by joinder, maturation, and growth of those cells was a variety of complex living organisms.

With the notable exception of Australia's lack of recent mountains, the geological history of various continents is generally similar. They all have sediments, mountains, rivers, plains, petroleum, volcanoes, etc. Their climates are generally similar when correlated with distances from the equator. If the conditions favorable for life to start were present on one continent they can be expected to have been present on other continents and some islands. The expected result would be a variety of organisms on each continent, basically similar, but with differences, mostly minor, some major. That is what is observed. The general pattern is similar. There are representatives of grasses, bushes, flowers, trees, birds, reptiles, cats, dogs, and monkeys on different land masses. There are almost invariably minor differences in specific members of flora and fauna of various continents. Even the grasses and trees are different. There are also marked differences such as the African male lion with its unique neck hair and claw-like structure at the end of his tail, the American pronghorn "antelope" with its unique horns, the Antarctic penguin with flippers, and the African giraffe, Australian platypus, and American armadillo with their obvious major distinctions.

Detailed evidence and reasoning to support the theory will be presented. First, the formation of petroleum will be presented. The reasons that it is probably not of biologic origin and the reasons that it is probably of inorganic, gaseous origin will be submitted. A discussion of chromosomes and of the chemicals of life, particularly deoxyribonucleic acid (DNA) and protein will be presented. Observations of plants and animals on various continents and islands that anyone can verify will be explained related to the presented theory.

II. THEORIES OF PETROLEUM ORIGIN

The prominent communication media, television, magazines, newspapers, and school textbooks, would have one believe that it is a proven fact that petroleum is a "fossil fuel." The picture is commonly presented that dinosaurs and other prehistoric animals "melted down" and became petroleum. Probably the main reason that people have accepted that idea is a rather superficial one. They have observed fats on animals, and it is apparently easy to accept the idea that those fats could become petroleum. Also, by the same type of superficial presentation, it is rather easy to convey the concept that the process of converting trees and ferns to coal if continued would produce petroleum.

Contrary to the appearance of only one viewpoint as to the origin of petroleum, its origin and the mechanism of its production are undetermined and subject of various theories. The simplistic ideas of fats and coal forming petroleum do not hold up under careful scrutiny.

There are some scientists who are convinced that petroleum is of organic or biologic origin, some maintain that it is most likely of inorganic or abiotic origin, and it has been proposed that some comes from both sources. It has been observed that usually geologists tend to favor the organic biologic theory of origin while chemists who have studied the problem believe that it is more probably correct that the origin of petroleum is from inorganic non-biologic sources.

The question of the origin of petroleum involves various scientific specialties. In order to explain petroleum production from formerly living organisms, its geologist proponents have resorted to postulating that anerobic bacteria act on the dead organisms and convert them to petroleum. There is actually no evidence of any bacteria having such capabilities nor evidence of bacteria in the depths of the earth. Geologists also postulate that chemical reactions occurred under conditions in which they are unlikely to occur.

The most obvious criticisms to organic biologic origin of petroleum will be considered first. The ordinary process that a log on the ground undergoes is to rot. Every schoolchild knows that when a log burns that the complex carbon-hydrogen molecules are converted by oxygen to carbon dioxide, CO_2, and water, H_2O. Rotting is a "slow burn." The final result of rotting is carbon dioxide, water, and some residual minerals or ash. The log forms neither coal nor petroleum.

Similarly, when an animal dies, all of its soft tissues "rot." As an intermediate process, enzymes within cells autolyze or digest the cells and cause the body to disintegrate. The end products of the carbon and hydrogen in fats, proteins, and carbohydrates are carbon dioxide and water. Only the minerals, mainly of the bones, are left as residue, and not a drop of oil of any type remains.

The tremendous volume of petroleum known to exist is an associated major objection to the organic, biologic origin theory. Known petroleum deposits or reserves have been estimated to be at least 600,000,000,000 barrels. The United Stated produced through 1978 approximately 117,800,000,000 barrels of oil. The area of the continental United States is 3,612,017 square miles. Therefore, the United States has produced through 1978 approximately 33,000 barrels of petroleum per square mile, or 0.005 barrels, per square foot. Adding the proven reserves of 27,051,289,000 barrels makes approximately an average of a quart of oil per square foot of surface of the United States.

Identifiable fossils are not common enough to account for such a huge amount of oil from similar fossil sources. On the other hand, there are more than ample amounts of hydrocarbons in the substance of the earth to form many times the known amount of petroleum. It has been calculated by comparison with amounts of hydrocarbons in meteorites that with 10% efficiency there are sufficient hydrocarbons in the substance of the earth to cover it to a depth of 40 meters with petroleum. The organic biologic theory must provide an explanation for conversion of living organisms to petroleum without their rotting to carbon dioxide and water. It is subject to serious question that any plausible biologic origin process can account for such huge quantities of petroleum that are known to exist.

III. BASIC NATURE OF PETROLEUM

Petroleum consists of hydrocarbon oils and waxes and is accompanied by hydrocarbon gases. Three types of chemical compounds are present, paraffins, cycloparaffins which are also called naphthenes, and aromatics. They may be found in any variety of proportions in different wells.

Paraffins are hydrocarbons in which the carbons are arranged in chains as opposed to rings or circles. They may be "normal" (n) consisting of carbons arranged in one straight chain, or they may be branched. Compounds are "isomers" if they have the same number of carbons and hydrogens but differ in how they are arranged, such as with branching. Examples of paraffin compounds are depicted structurally in *Figure 1.*

Paraffin hydrocarbon gases found with petroleum include those with one through four carbon atoms in a chain in one molecule. The liquid and waxy paraffin compounds in petroleum include all those having five or more carbon atoms in one molecule.

Cycloparaffins or naphthenes are hydrocarbons in which the carbons are arranged in rings of five or six carbons in each ring. They are "saturated" with hydrogen. That means that the usual four bonds of carbon are single bonds and there are no double bonds between the carbons where more hydrogens could be added. Examples of small cycloparaffins are depicted in *Figure 2.*

The cycloparaffins are generally heavier, having more carbon and hydrogen atoms per molecule, than the paraffins. They are an important part of the residues after the lighter oils such as gasoline and diesel fuel are boiled off and separated by distillation. The ring compounds may consist of numbers of rings, up to five rings in a molecule having been identified.

The aromatic hydrocarbons have "unsaturated" carbon rings with "resonating" double bonds between carbons. They are called "unsaturated" because at sites of double bonds other atoms such as hydrogen can be added. Each carbon can hold four other atoms by single bonds. The structure of benzene, an aromatic hydrocarbon, is depicted in *Figure 3.*

The aromatic compounds comprise only a small percentage of most crude oils. Of that small percentage, mixed aromatic-napthenic hydrocarbons are most common. That means that most aromatic rings are found joined to saturated naphthenic rings.

No unsaturated chain hydrocarbons are found in petroleum.

Paraffins can be found attached to any of the rings thereby making the resulting compound larger, more complex, and heavi-

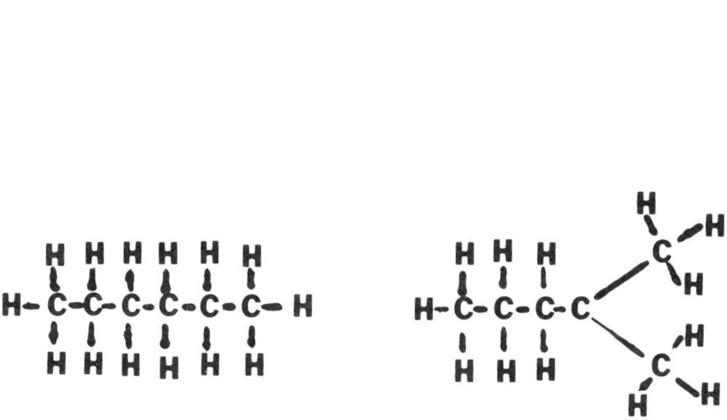

n–hexane isohexane

Figure 1. Examples of paraffin hydrocarbons, *n*–hexane and isohexane, two isomers of hexane.

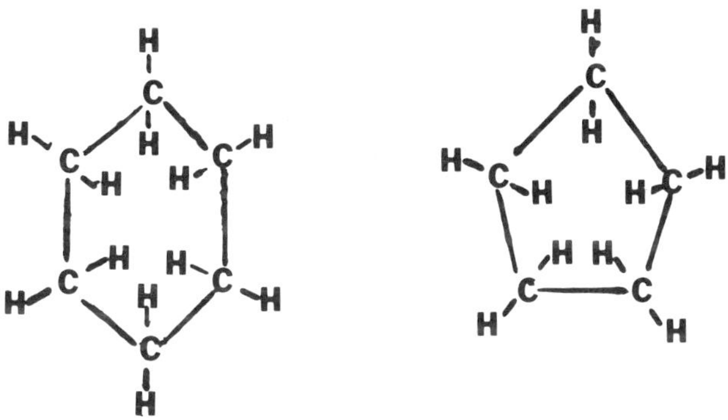

cyclohexane cyclopentane

Figure 2. Examples of cycloparaffins.

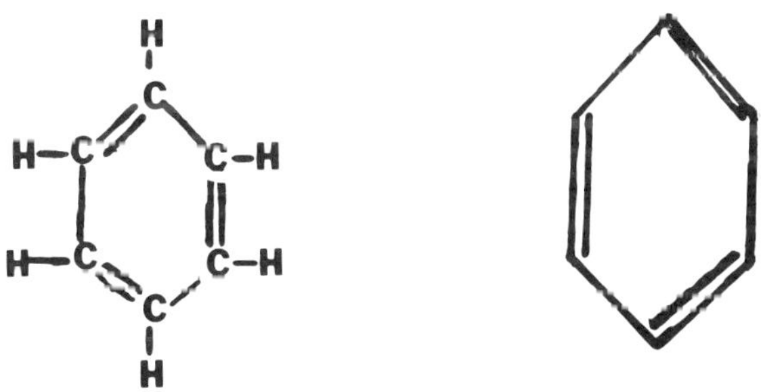

Figure 3. Benzene, an aromatic hydrocarbon. On the right the structure is shown in the common abbreviated manner with carbon atoms and hydrogen atoms not designated by letters.

er. Further, in addition to pure hydrocarbons, atoms other than hydrogen or carbon can be found attached to paraffins, cycloparaffins, or aromatic hydrocarbons. Sulfur, nitrogen, and oxygen are elements commonly found in such compounds. It has been claimed that petroleum may include many thousands of different hydrocarbon compounds. Most have not been completely identified as to their precise formulae and structures.

In spite of variations in proportions of exact types of hydrocarbons in different wells, a general uniformity of chemical composition of petroleum from different sources is recognized. They all contain similar hydrocarbons. Almost all crude oils contain between 83 and 87 per cent carbon, 11 to 14 per cent hydrogen, and 2 to 3 per cent nitrogen and sulfur. The uniformity is greater than expected in the organic theory. Petroleum is found in sediments or formations of essentially all geologic periods. The organisms living in each of those time periods were different. Local conditions such as types of clay, chemicals, heat, and pressure to organisms undoubtedly varied in those different periods. The uniformity therefore appears to be inconsistent with the organic biologic theory of origin of petroleum.

The general uniformity of petroleum is not inconsistent with the inorganic origin theory. The gases from magma combined to form chemical compounds. The conditions and gases involved in those chemical reactions were essentially the same wherever petroleum was produced.

IV. GASES FOUND WITH PETROLEUM

There are three general substances found in petroleum wells. There are gases, liquid petroleum, and water in which there are dissolved minerals. Examination of the characteristics of the three parts reveal that the organic biologic theory of petroleum origin does not satisfactorily explain either the major or exceptional features of any of those general substances.

The most common hydrocarbon gas is methane, CH_4. Methane can be found in wells with or without significant quantities of liquid petroleum being present. Other hydrocarbon gases commonly found with petroleum are ethane, C_2H_6, propane, C_3H_8, and butane, C_4H_{10}.

The presence of methane is usually considered evidence for the organic theory because methane is the common "swamp gas" observed as the result of decomposing organic material. Its presence with petroleum, as will be explained, can also easily be accounted for by the abiotic, inorganic theory.

In considering methane as evidence supporting the biologic organic theory, the presence of the gases of larger hydrocarbons seems usually to be ignored. Ethane, propane, and butane are not found in living organisms. They have not been demonstrated to be products of bacterial action or natural decomposition of dead organisms.

An apparent possible explanation for the presence of the larger gas molecules, ethane, propane, and butane, would be that the molecules of methane join together, a process called polymerization, to form larger molecules. That process is represented in the following equations.

CH_4 (methane)	+	CH_4	$\longrightarrow$	C_2H_6 (ethane)	+	H_2 (hydrogen)
C_2H_6 (ethane)	+	CH_4	$\longrightarrow$ yields	C_3H_8 (propane)	+	H_2

In commonly proposed theories of organic biologic origin of petroleum there are no situations where such polymerizations could be expected to occur. In the biologic theory, complex chemicals of formerly living organisms containing hydrogen and carbon are claimed to have been broken down to form hydrocarbons, and not the reverse of small molecules being built into larger molecules. If polymerizations of methane had occurred, hydrogen gas would also have been produced as shown in the above equations. Hydrogen gas, H_2, is absent or occurs only in trace amounts in petroleum wells. That proves that in the organic theory of petroleum origin, ethane, propane, and butane could not have been the result of polymerizations of methane. As they cannot be accounted for by decomposition of organic matter, their presence with petroleum is simply unaccounted for in the biologic organic theory of origin of petroleum.

By way of contrast, briefly, in the inorganic theory the complex hydrocarbons are the product of polymerizations of carbon atoms and hydrogen atoms in situations where there are no excess hydrogen atoms. For that reason no hydrogen gas is formed.

Gases which are not hydrocarbons are found not uncommonly with petroleum or with hydrocarbon gases. They include helium and argon. Those particular gases are never found in living organisms. They are never found to result from decomposition of living organisms.

In the biologic petroleum origin theory, the presences of helium and and argon with petroleum are explained as having been produced independently from the formation of petroleum. Helium is usually accounted for as being a product of radioactive decay of minerals such as uranium, being ejected from the nuclei of such heavy radioactive elements. It is believed by proponents of the organic theory of petroleum that it is coincidental that helium and hydrocarbons are found in the same trap. Argon is postulated as having resulted from radioactive potassium. It has also been postulated that the helium has come from basement rocks along with nitrogen.

Helium is found in concentrations up to 16% of total gases in a natural gas well. In the organic theory of petroleum formation, with helium being only coincidentally in the same trap as methane and petroleum, one might expect to find helium in some wells without methane present. Since it is not found without methane being present, that is submitted as evidence that helium is not due to some process unrelated to the presence of methane. Because helium could not have come from organic matter with methane, a satisfactory explanation for its presence by that theory is lacking.

In the inorganic theory, methane, helium, argon, and other gases from the Earth's interior and substance can be expected to be found in various concentrations and with other gases that commonly come from that same source.

Nitrogen may be present in concentrations up to 96% of the natural gas from wells. Decomposition of living organisms can be expected to yield nitrogen as ammonia, NH_3, but not as nitrogen gas, N_2. There is no way the biologic organic theory of petroleum formation can account for such high concentrations of nitrogen. Other gases would necessarily also be present.

In the inorganic theory the source of nitrogen found with petroleum is the same as its source was for the atmosphere. That gas which makes up approximately 80% of the gas of the atmosphere is generally believed to have come from the hot interior of the Earth.

Carbon dioxide gas is commonly found with petroleum. Approximately 0.5% of the natural gas of commerce is carbon dioxide, CO_2. At the extreme, its concentration may be up to 99% of the gas from a well. Such a high concentration is incompatible with the organic biologic theory of petroleum formation. In addition to other gases necessarily produced, that much oxygen would have prohibited the production of petroleum in the theory. High concentrations of CO_2 are expected to occur at times in the inorganic theory. As will be presented in more detail later, carbon dioxide is commonly expelled as a gas released from the hot interior of the Earth through volcanoes and fumaroles.

It is apparent from the foregoing that analyses of the gases found in petroleum wells seriously erode the credibility of the biologic organic theory of petroleum formation. The presence of hydrocarbon gases larger than methane, namely ethane, propane, and butane, has not been satisfactorily explained. The theory does not account for the absence of hydrogen which would be produced if those gases were from biologic sources.

A scientific theory is stronger and more likely true if it can explain all relevant observed phenomena. The organic theory cannot satisfactorily explain the presences of helium and argon without utilizing theories of their production from inorganic sources.

The organic biologic theory fails to account for the extreme of gas concentrations found. It simply cannot account for concentrations of nitrogen of 96% and carbon dioxide of 99% which have been found.

The absence of hydrogen gas is generally recognized as a defect in the biologic origin of petroleum. In addition to hydrogen being expected from polymerizations of methane, alpha radiation, which is expected in some of the proposed source beds for petroleum and in routes travelled, produces hydrogen when it acts on organic matter. If petroleum were the product of decay of biologic material, free hydrogen gas would be present. To the contrary, it is not found except perhaps in traces.

V. OIL FIELD WATER

The presence of water with petroleum and the minerals dissolved in the water further damage the credibility of the biologic theory of petroleum formation. As noted above, when a formerly living organism rots, the carbon and hydrogen making up its substance combine with oxygen and form carbon dioxide and water. A key element is oxygen. When oxygen is present in amounts sufficient to produce water, carbon dioxide, CO_2, is formed, the organism rots, and methane and petroleum cannot be formed. In the organic theory of petroleum formation it is essential that organisms be buried and converted to petroleum in anerobic (no oxygen) or nearly anerobic conditions. The formation of water which requires oxygen is completely inconsistent with that process. The formation of water would be only consistent with the formation of carbon dioxide from the dead organisms and not methane and other hydrocarbons.

It was noted above that to account for the presence of helium, and argon, the proponents of the organic theory have explained the presence of those gases by two additional hypothetical processes. To explain the presence of water with petroleum, another independent process has been proposed. It has been theorized that the water was initially in the sands before the petroleum arrived. The postulate requires that petroleum found with water has always been produced at some site different from where it is now found. The theory is that after it was formed at some distant site that it migrated into a geologic formation where water was already present.

In order to logically consider the significance and probable origin of water which is found with petroleum it is pertinent to examine the origin of other water. The volume of the oceans has been estimated to be eleven times the volume of the land mass above sea level. It is generally accepted that the major source of the water of oceans was probably from gases released from within the body of the Earth through vents such as volcanoes. It has also been proposed that other substances including carbon dioxide, chlorine, sulfur, and nitrogen, have also come from that source particularly because of their abundances. The salt dissolved in the oceans has been estimated to be at least 4,500,000 cubic miles, or about 14.5 times the bulk of the continent of Europe above sea level. Its composition is 78% sodium chloride, 11% magnesium chloride, 4.8% magnesium sulfate, 3.6% calcium sulfate, 2.5% potassium chloride, 0.22% magnesium bromide, and 0.34% calcium carbonate.

Any of the elements in those chemical compounds may be found as residue from organisms that have decayed or rotted. But that is probably not the initial source of those elements in the ocean. Those elements in the ocean probably came from the Earth. Water and essentially all of those elements and others are still being emitted from within the Earth through volcanoes and fumaroles.

The water found with petroleum is usually a very concentrated brine. It may be as much as 17% salt. By contrast, sea water is about 3.4% salt. Chloride concentrations may be 2 to 8 times that of sea water. The calcium-sodium ratio may be higher than in sea water. Magnesium is found in a concentration lower than that of sea water and sulfates may be scarce. With magnesium and sulfates in lower concentration than in sea water, the concentration of the petroleum well brine is not simply a matter of removal of water from trapped sea water.

Fatty acids including naphthenic fatty acids may also be present in the water with petroleum. The presence of fatty acids can be expected regardless of whether the petroleum is from biologic or inorganic sources. The absence of phosphates, carbohydrates, amino acids, proteins, or other soluble complex chemicals of living organisms is difficult to explain if the fatty acids are there and unchanged from formerly living organisms. As will become apparent in the discussion of DNA to follow, phosphates are essential to life. They are common in all living organisms. Their absence in petroleum and in water with petroleum seriously erodes the credibility of the organic biologic theory of petroleum formation. Naphthenic fatty acids are rare in biologic matter, and their presence in relatively large amounts compared to other fatty acids present in oil field water is unexplainable in the biologic organic theory of petroleum formation.

The presence of water with petroleum and analyses of the contents of the water damage the credibility of the organic biologic theory of the origin of petroleum. In the organic theory the production of water with petroleum would annihilate the postulate of anerobic conditions essential to that theory. Therefore, a separate origin of the water has been postulated along with the proposition that when water is found that the petroleum was always formed somewhere else and migrated to where it is now found. The mineral content and fatty acids found in the water with petroleum are inconsistent with the organic biologic theory.

In the inorganic abiotic theory of petroleum formation the presence of water is expected. This theory postulates that the oil field water with its minerals was formed from the same source as the petroleum, from gases from within the Earth. The occurrence together of the water, its dissolved minerals, the gases present, and the petroleum are consistent with their being the result of gases expelled from within the hot Earth.

VI. PORPHYRINS

Findings of porphyrins in petroleum have been considered to be essentially conclusive evidence of production of petroleum from biologic sources. Porphyrins are carbon-hydrogen compounds with a basic structure depicted in *Figure 4*. Chlorophyll is a magnesium-porphyrin. *Figure 5*. Hematin is an iron-porphyrin which is an essential part of hemoglobin. It has been claimed that porphyrins in petroleum came from chlorophyll and possibly hemoglobin

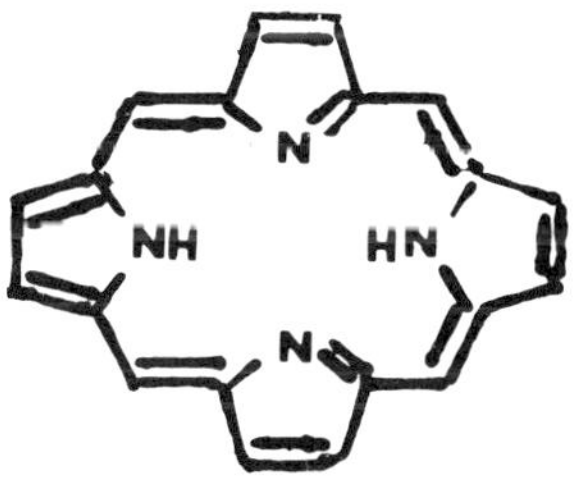

Figure 4. The basic struture of porphyrins.

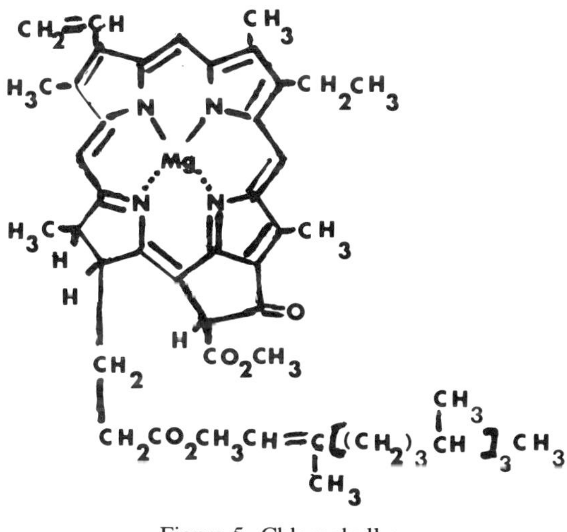

Figure 5. Chlorophyll *a*.

Neither chlorophyll nor hematin are the porphyrins found in petroleum. The porphyrins found have been nickel-porphyrin compounds and vanadium-porphyrin compounds. As for the nickel compounds, it has been postulated that hematin plus nickel salt yields a nickel-porphyrin compound which is more stable than the iron-porphyrin compound. Even if that reaction could occur, which is questioned, it would not account for no hematin being found when no nickel is present. Further, that reaction would not account for no chlorophyll being found if the source was formerly living organisms.

Usually whether one metal will displace another can be predicted from its position in the electromotive series of elements. In the electromotive series, the pertinent voltages are: magnesium 2.37, vanadium 1.18, iron 0.44, nickel 0.25. That series means that magnesium would be expected to displace the other three in compounds with other elements. Vanadium would be expected to displace iron and nickel, and iron would be expected to displace nickel. If chlorophyll, a magnesium-porphyrin compound, was a source material for petroleum, and porphyrins survived the conversion to petroleum, chlorophyll would be expected to be found in petroleum. Neither vanadium nor nickel would be expected to displace magnesium. One might expect vanadium to displace the iron in hematin. Nickel would not be expected to displace iron or magnesium. From the foregoing, it is plain that the finding of nickel-porphyrin is inconsistent with the organic theory of petroleum formation. The lack of chloryphyll when there are other porphyrins present in petroleum is inconsistent with the postulate that porphyrins found in petroleum are from plants.

Mainly because porphyrins can be expected to occur along with the many other hydrogen-carbon-nitrogen compounds in petroleum in the inorganic abiotic theory of origin, their presence is probably due to their being chemicals formed along with other complex hydrocarbons comprising petroleum. Confirming that probability, it has been reported that vanadium porphyrins can be produced from igneus rocks. Also, porphyrins can be produced in laboratories from ammonia, methane, and water.

The proponents of the organic theory claim that porphyrins passed from the organisms in which they were formed into petroleum without structural change. To be consistent, those conditions restrict the entire petroleum production process from start to finish to conditions that preserve porphyrins.

Porphyrins break down at temperatures above 200 to 250 degrees Celsius. Therefore, in the organic theory the temperature could not have exceeded 250 degrees throughout the history of petroleum formation, migration, and storage. In the inorganic theory there is no problem since the porphyrins were formed during the cooling phase.

Those temperature restrictions are incompatible with the chemical reactions required to produce petroleum. Aromatic compounds found in petroleum cannot be formed at less than 500 degrees Celsius. Below that temperature, the reactions result in quite different compounds.

The temperature of petroleum formation is restricted even more than that in the biologic theory of origin. According to that theory, in the laying down of ocean sediments, in the migration of the liquid petroleum which was formed, and during its storage it was not exposed to significant heat. Petroleum in some shallow wells has not been exposed to more than 40 degrees Celsius according to the biologic origin theory. Some regard 60 degrees Celcius as the upper temperature limit in the history of most petroleum formation. As noted above, those low temperatures are inconsistent with the temperatures required to form the aromatic hydrocarbons commonly found in petroleum.

It has been suggested that high pressures without high temperatures could have converted organic matter to petroleum and particularly with catalysts present. That postulate also encounters problems.

The proposal requires pressure to cause changes from animal and/or plant materials to gases and other hydrocarbons. The end products would have greater volume and occupy more space than the initial material. The effect of pressure must be to produce a decreased volume, not the opposite as postulated. Further, in laboratory experiments pressures up to 15,000 times atmospheric pressure have failed to produce certain important petroleum compounds from postulated precursors.

Another serious objection to that postulated role of pressure is that in the organic theory of petroleum formation there are no conditions where a sustained high pressure is expected. In that theory, petroleum is believed to have been produced in source beds or in route to sites where it is presently found. That required production at a site from which the petroleum could move. It was not a closed system where high pressures could have been maintained. It is generally proposed that dead organisms were laid down in fine-grained sediments. Pressure upon those sediments is believed to have been the prime force causing movement from that sediment to coarser material. One might expect that squeezing process to have caused fats and oils in the organisms to have migrated without being changed to petroleum hydrocarbons. Such unchanged fats and oils are not found in petroleum. Once the petroleum has migrated into sands covered by hard capping rock, there is still no reason to expect consistently high pressures. Water transmits its pressure, but pressure is only minimally transmitted through solids. Most petroleum wells are not "gushers." Most petroleum is not found to be under pressure. It simply occupies the space between grains of sand or rock or between the grains of sand of which sedimentary rock is composed.

The finding of porphyrins in petroleum cannot be considered as evidence in favor of the organic theory of petroleum formation. They are the wrong porphyrins. Porphyrins found are vanadium and nickel porphyrins, while chlorophyll is a magnesium porphyrin and hematin is an iron porphyrin. It is unlikely that nickel porphyrins were produced from either chlorophyll or hematin. The finding of porphyrins and not chlorophyll is inconsistent with chlorophyll being their source in petroleum. The presence of porphyrins is not unexpected in the inorganic theory where thousands of chemical compounds comprising petroleum were formed from simple elements. Further, porphyrins have been produced from rocks of volcanic origin and in laboratories from methane, ammonia, and water.

The finding of porphyrins in petroleum is evidence of the new theory presented here of the origins of the chemicals of life. It is evidence that whatever was required to produce the porphyrin rings of chlorophyll and hematin was present during petroleum formation.

VII. OPTICAL ACTIVITY

Optical activity of petroleum is commonly considered together with the presence of porphryns to be the deciding or conclusive evidence that petroleum came from previously living organisms. As a matter of history, when the organic theory of petroleum origin was first formulated about 1900 A.D., optical activity was considered to be characteristic of life's chemicals and without life it was believed that there could be no optical activity of chemicals. Optical activity is more fully understood now. It is not proof of biologic origin.

A substance is said to be optically active if it causes the plane of polarized light to be rotated. A beam of polarized light is passed through a solution. If the plane of the light is twisted clockwise, it is dextra-rotatory. If it is twisted counter-clockwise, it is levo-rotatory. Tested petroleum samples have caused polarized light to be very slightly rotated up to 1.2 degrees. The majority of the samples tested were dextra-rotatory, some were levo-rotatory, and others were optically inactive. The optically active compounds in petroleum have not been specifically identified. They are present only in fractions boiling between 250 and 300 degrees Celsius.

The biologic theory of origin of petroleum fails to account for lack of optical activity in all fractions. Also, if formerly living organisms were responsible for the optical activity, with presently accepted concepts all petroleums would be expected to rotate the light in the same direction which is not the case.

A further examination of principles involved in optical activity is required to understand its important role in the inorganic abiotic theory of petroleum formation and the origins of the chemicals of life. If two molecules differ only in that one is the mirror image of the other and they cannot be superimposed on each other, they are asymmetric. They are "right-handed" or "left-handed." They are different like a right glove is different from the left. They cannot be superimposed. That would be like trying to put a right hand in a left glove. Asymmetric compounds are optically active.

Optical activity is common with carbon compounds. Any compound having a carbon with four different elements or combinations of elements attached to it is asymmetric and thereby optically active. For example, the hydrocarbon compound depicted in *Figure 6* is asymmetric and is either dextra-rotatory or levo-rotatory.

A mixture of equal parts of the dextra and levo compounds is called a racemic mixture. It is optically inactive because the rotations cancel each other out. In the usual laboratory chemical reaction, such a racemic mixture is the usual and expected result. In order to obtain an optically active substance, one group, the D or L, must be suppressed in formation, destroyed, or separated in some way from the mixture.

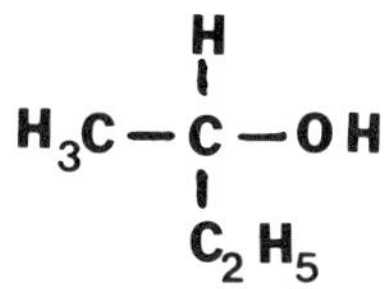

Figure 6. Because the carbon atom in the center has four different groups attached to it, it is asymmetric, right-handed or left-handed, and optically active.

Catalysts are substances such as chemicals or enzymes which do not enter a chemical reaction but enhance it or cause the reaction to occur. An optically active catalyst will result in the production of compounds having the same direction of optical activity. In the presence of such an optically active catalyst, the D (dextra-rotatory) and L (levo-rotatory) forms probably react at different rates to the extent that from a practical result one doesn't react. That is a means of producing an optical active compound. Also, the D and L forms may have different solubilities particularly in an optically active solvent. That can be used to separate them from a racemic mixture. Enzymes or bacteria may act selectively upon D or L forms and effectively separate them. Those types of situations are examples of optical activity breeding other optically active substances. Those were the types of observations that formed the basis for optical activity being considered to be proof of the organic theory of petroleum formation.

Optically active substances have been produced without utilizing optically active agents. Exposure to circularly polarized light can produce that effect on a chemical reaction. Sunlight reflected from the surface of the water reacted upon by the Earth's magnetic field can be a source of such circularly polarized light. Radioactivity can decompose one of the rotatory forms faster than the other. As an example, d-tyrosine is destroyed faster than l-tyrosine by the effect of radioactive decomposition of strontium-90. A magnetic field such as that of the Earth can cause any substance to exhibit optical activity with the resulting direction of rotation depending upon the direction of the magnetic lines of force.

Pasteur utilized a simple system to separate optically active forms from a racemic mixture. He utilized the principle that a racemic mixture may crystallize in such a manner that all of the dextra-rotatory molecules go into one crystal and all of the levo-rotatory molecules go into another. One may crystallize before the other and then be separated by filtration. Spontaneous crystallizations of one active form, the D or L, have been reported. That could separate the D or L form and breed other optically active compounds.

The molecule of a chemical compound does not need to be asymmetric to form a crystal which is asymmetric. Examples of such include silicon dioxide and particularly quartz crystals. An asymmetric quartz crystal can catalyze or breed optically active chemicals. One would expect that in nature an equal number of dextra-rotatory and levo-rotatory quartz crystals would occur. Actually, right-handed quartz crystals are considerably more common in nature than left-handed ones.

Effects of quartz crystals, spontaneous crystallizations, radiation, circularly polarized light, and magnetism are examples of production of optically active compounds without biologic matter. The effects of any one or more of those might account for the slight optical activity of petroleum.

Studies of optical activity of biologic matter are important in understanding origins of life as well as origin of petroleum. Proteins are composed of only L-amino acids. The sugars of nucleic acids are only D-ribose and D-deoxyribose. The designations of D and L to amino acids have been based upon their configurations rather than the actual direction they cause light to be rotated. L-amino acids apparently rotate light to the right. That is attributed to the effect of side chains.

Pauling and Corey found that most proteins are in the configurations of an alpha helix. Wald did further experimental work on the structure of proteins and optical activity. Proteins in alpha helix structure, a corkscrew-like configuration, cause polarized light to be rotated to the right (clockwise) 48 degrees. Again, the maximum rotation caused by petroleum was only 1.2 degrees. L-amino acids were found to contribute to the dextra rotation, but it was mainly due to the helix. The right-handed alpha helix prefers L-amino acids. The three dimensional configuration allows for better linkage of L-amino acids in a right-handed helix and of D-amino acids in a left-handed helix. The atoms linking together fall closer together if amino acids are either all left-handed or all right-handed. When D-amino acid was the only one available in an experiment, the helix started out to the right but then reversed direction and produced a left-handed helix corresponding to the abnormal amino acids. With a mixture of D and L amino acids the helix propagated in the normal right-handed direction but it grew more slowly and was less stable than when only the L form of amino acid was used. The result from the mixture of D and L amino acids was a helix weighing only one-fourth as much as a helix constructed with only the L amino acids available as in nature.

Nucleic acids, deoxyribonucleic acid (DNA) and ribonucleic acid (RNA), will be treated more thoroughly later, but are important here in this context. Nucleic acids have a similar right-handed helical configuration with two strands side by side joined together. The three dimensional shapes of the sugars in the nucleic acids require them to be of either all L or all D form in order for the right atoms to fall close enough to each other to form the bonds between the strands. The right-handed helix requires the sugars to be the D form. If the helix were left-handed, then the sugars would have to be of the L form. Or, looking from another direction, D sugars will result in a right-handed helix form of nucleic acid. Based on the protein studies, L sugars would probably result in a left-handed helix. The alpha or right-handed helix of DNA is schematically depicted in *Figure 7.*

Wald, Miller, and others have assumed that it was a 50-50 chance that the first living organism happened to construct its proteins with L-amino acids. They have postulated that when the first living organism was constructed with only L-amino acids it determined that all organisms following would have the same make-up regardless of whether or not that first organism was ancestor to all that followed. It would have established the type of nutrients available. (It may be noted parenthetically that proponents of the theory of evolution could have also reasoned that other organisms would have developed to fill the niche and use the D-amino acids left behind.) The same 50-50 chance reasoning can be applied to use of only D-deoxyribose to build DNA or D-ribose to build RNA. At this time it is impossible to know which was formed first, proteins, DNA, or RNA.

It might have been a 50-50 chance for use of D or L chemicals as proposed if this were a flat, non-moving world. We live in an asymmetric world. The sun spins in one direction. That is asymmetrical. As the solar system moves through space each of the planets travels in a helical path. Some galaxies are helical. Every helix is asymmetric. It cannot be superimposed upon its mirror image. It is either right-handed or left-handed. The Earth is not symmetrical in any respect. The land and sea are not arranged symmetrically or in even a stable configuration. The Earth is not symmetrically round, but is pear-shaped. The Earth spins in one direction. That direction is oppositely oriented to persons on opposite sides of the equator. Electric currents and magnetism are asymmetric. The magnetic lines of force go in one direction oriented to the flow of electricity through a wire. If electricity is passed northward through an electric wire oriented north and south, the needle of a compass lying over the wire points to the west. Gravity acts to pull in the direction of the Earth with resulting differences of up and down an asymmetry.

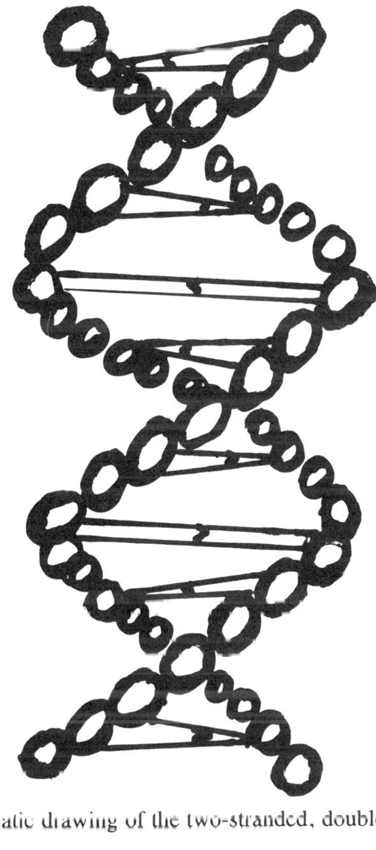

Figure 7. A schematic drawing of the two-stranded, double helix configuration of DNA.

Helices are common in nature. Plants exhibit them in arrangements of leaves and in twisting of stems. Mollusc shells are helical. In the human the umbilical cord is twisted in a helix. The heart is on the left, the liver is on the right, the spleen is on the left. The intestines are rotated. Those asymmetries arise during development of the embryo by suppression of growth of opposite structures and by rotations of developing organs. The left side of the brain in most persons is dominant with resulting right-handedness. Young female birds each have an ovary and oviduct, or tube, on both right and left sides. The right tube and ovary degenerate and the functioning left tube becomes greatly enlarged to allow the passage of eggs.

With such prevalent asymmetries there is no reason to assume that selection of asymmetric chemicals to build living organisms was a 50-50 chance just because a cause is not readily apparent. It has generally been assumed that the chance picking of L-amino acids for proteins and D-sugars for nucleic acids determined the direction of the helices.

Another explanation is proposed based upon forces of the rotating Earth and gravity. Because of the rotation of the Earth, ocean currents, tornadoes, typhoons and hurricanes rotate in opposite directions in Northern and Southern Hemispheres. In 1902, Ekman determined that the relevant forces from the surface to the bottom of the ocean in ocean currents is a logarithmic spiral. The spiral effect is easily observed in whirlwinds and tornadoes. The effect can also be demonstrated in water swirling out of sinks, although the direction of the swirl may be affected by the way water entered the sink and irregularities in the sink and in the opening of the drain.

The movement of ocean currents results from inertia of the water. As the Earth rotates from east to west, the water lags because of its inertia. The rotating movement requires acceleration, and the water tends to move in a straight line. The lagging of the water makes it move from west to east in relation to the earth below it. Actually, of course, it is the Earth as a whole that is moving faster than the water. With each rotation of the Earth, an object at the equator moves approximately 25,000 miles, or about 1,000 miles per hour. At the opposite extreme, an object at the north pole does not move any distance but just rotates. Because the surface of the Earth moves more rapidly near the equator, predominant ocean currents near the equator move more rapidly than those farther from the equator. In the Northern Hemisphere, the predominant ocean current near the equator flows from west to east and as it approaches land it turns the direction of least resistance toward the north pole. The result is the clockwise current.

Predominant ocean currents in the Northern Hemisphere rotate clockwise. Water flowing from a sink in the Northern Hemisphere usually moves counterclockwise. The winds of hurricanes in the Northern Hemisphere move counterclockwise. Most tornadoes in the Northern Hemisphere move counterclockwise. All of their counterparts usually move in the opposite directions in the Southern Hemisphere.

In the construction of a helical protein molecule or a helical molecule of DNA, it is postulated that as molecules approach the larger developing mass, the direction of their approach is altered by the rotation of the Earth just as water swirls in ocean currents and just as particles swirl in a draining sink. As their direction of approach is twisted, and each molecule is twisting, each successive molecule that joins tends to be deposited in the same swirling helical pattern. The resulting conglomerate of molecules forms a helix. A helical structure conforming to those gravitational influences would be expected to be more stable than a non-conforming structure. It may be recalled from above, in Wald's experiments when only D-amino acids were available they start to build a normal right-handed alpha helix but switched directions and formed a left-handed helix to accommodate the configuration of the D-amino acid building materials which were opposite in optical activity to those of nature. The initial choice of direction of the helix was normal and opposite to the one dictated by the availability of only D-amino acids. The choice of direction of the helix is postulated as having been initially directed by the rotation of the Earth and gravity. The resulting large protein molecule has a clockwise, right-handed, alpha helix configuration. The selection of D or L forms of chemicals used is postulated as having been determined by the direction of the helix when either D or L materials were available. Protein helices required L-amino acids for their most stable molecules. DNA and RNA alpha helices could only use D-sugars. Whichever of those helices that formed first could have determined the direction of the others or they may have each started independently. The foregoing accounts for optical activity of biologic matter.

If the foregoing hypothesis is correct, the implications are boundless. For example, there could be right-handed and left-handed proteins and DNA depending upon in which hemisphere it was formed. It could have something to do with the vast differences in flora and fauna of the Northern and Southern Hemispheres. Any plant or animal living in the Northern or Southern Hemisphere that did not originate there would not be in ideal balance with nature. A shift of the north and south poles could cause the same result. The D or L components of protein not in conformity with the helix of the particular hemisphere in which it is located could be related to diseases of proteins and aging. There could be left-handed and right-handed viruses and bacteria with immunity affected. Our diets might need to be changed to better accommodate our structures. It may aid in determining the history of the axis of the Earth.

As for optical activity of petroleum, heavy fractions containing the larger molecules are the only ones in which optical activity has been demonstrated. Some of those larger molecules formed through polymerizations probably have helical shapes because of the same reason proteins and DNA formed helices. That is postulated as being the main cause of optical activity in petroleum. As presented above, other possibilities include effects of quartz, radiation, circularly polarized light, and magnetism. For certain, optical activity of petroleum is not proof of biologic origin.

VIII. CLAIMED ORGANIC SOURCES

Proponents of the theory that petroleum is from formerly living organisms have not reached agreement as to the exact sources. The most commonly proposed are coal, oil shale, fish, and the small algae and plankton.

Coal

It has not seemed completely unreasonable that people have believed that the process that produced coal from ferns and trees if continued would have produced petroleum. The color of coal, its occurrence underground, and its flamability suggest such a possible relationship with petroleum. Considerable efforts have been made to determine whether there is in fact such a relationship.

Geologic studies have not been fruitful in establishing a connection between formation of coal and petroleum. First, petroleum is not found in significant amounts with coal. When it is found with coal, it is usually in fissures or cracks where its presence can be accounted for by migration from some other source. Even when coal beds and oil fields are in the same area, they are totally distinct and separated so that it is clear that the oil has not come from the nearby coal beds. The largest deposits of petroleum are in strata of geologic periods which were not coal-forming. As noted, petroleum and natural gases can be found in essentially all geologic systems or periods from precambrian to recent. Coal deposits are restricted to much fewer geologic periods and strata and the lowest geologic layers in which coal is found are above layers in which petroleum is found. That indicates that petroleum was formed at earlier times than ferns and trees were deposited to form coal.

Also in conflict with coal being a source of petroleum, no material has been found which is intermediate between coal and petroleum. A relationship has been shown in the peat-lignite-bituminous-anthracite series in the formation of coal, but there is nothing corresponding in the direction of petroleum from any thing in that series or any other biologic organic material.

Chemically, coal is quite different from petroleum. The ratio of carbon to hydrogen is 6 to 7 in petroleum, 7 to 9 in shale, and 10 to 16 in coal.

From the foregoing, it is clear that scientific evidence refutes any claim that petroleum has come from coal.

Oil Shale

Shale is rock that is formed from clay or clay-like, fine-grained sediments, and is characterized by laminated layers forming cleavage planes. "Oil shale" is shale and associated clays which contain organic material, remnants of formerly living organisms. Its contents are also known as kerogen. Microscopic observations have revealed that the organic matter consists largely of macerated remains of plant life including algae, spores, pollen, and resins. It is generally agreed that the remains are mainly from vegetable life.

As the organic content of shales increases, the kerogen shales grade imperceptibly and gradually through "torbanites" and "boghead coals" into "cannel coal" or "kerogen coal." There is no claimed relationship between this type of coal and the bituminous series which we commonly know as coal, nor is there an intermediate product between kerogen and petroleum.

There is no real evidence by scientific standards that kerogen in nature forms petroleum. Migration can account for all of the occasional liquid hydrocarbon found in oil shale.

When kerogen is heated to 350 degrees Celsius it is decomposed into gas, oil, and residual carboniferous material. The oil produced from oil shale is different than petroleum. It has a higher percentage of nitrogen and oxygen than petroleum; and as noted above, the carbon-hydrogen ratio is also different.

Oil is produced from kerogen only by heat. There is no uniform source of heat of the degree required in the biologic origin theory. It has been proposed that volcanoes supplied the necessary heat. Many oil fields are located far from volcanoes which precludes that possibility. Furthermore, such heat would have destroyed the porphyrins, the presence of which is heavily relied upon by the proponents of the organic biologic theory of petroleum origin.

Heat required for conversion of oil shale to liquid oil, chemical differences in petroleum and oil produced from shale, and the lack of geologic association of shale and petroleum fields prove that oil shale was not a significant source of petroleum.

Fish

It is common knowledge that fish contain considerable oil. As might be expected, it has been claimed that they provided the main source of material for petroleum production. Other than the superficial observation of oil content of fish, the evidence refutes the claim.

The fish-origin theory requires combinations of events and situations which have extremely low probability of having occurred. To prevent ordinary decay and rotting, the fish would have needed to have died in huge numbers at nearly the same time and to have been buried quickly. That would have required cataclysmic deaths of fish in most parts of the world in each of the geological periods. Fossil and other evidence of such having occurred are lacking.

It has been proposed that volcanic activity could have killed the fish, roasted them to produce oil, and then started its migration. Evidence refutes that idea. Petroleum is found in geological strata which were established before the fish were present. No residues from heat acting on fish or any other organic material have been found in relationship to any petroleum field. It is recognized that petroleum is not found in significant amounts near volcanoes.

In addition, the heating of fish oil could no more be expected to yield petroleum than the heating of oil shales. The resulting product might resemble petroleum in appearance, but chemical analyses would prove them to be quite different. Furthermore, carbon isotope studies rule out an animal origin of petroleum.

Algae, Plankton

Mainly because of the failure of all other biologic material to account for the production of petroleum, the proponents of the organic biologic theory of its origin have speculated that the most likely source material was small plants and animals such as algae and plankton. There seems to be little question that deposits of them could have produced oil shale; but there is no evidence that they were the source of petroleum. The theory requires that some deposits of algae and plankton resulted in oil shale while other similar deposits formed petroleum.

Because oil shale cannot be converted to petroleum, it has been hypothesized that bacteria acting on the algae and plankton produced petroleum. There is no evidence that bacteria having such capabilities exist. Over two thousand samples of ocean bottom sediments have been studied. Organic matter has been found, but no liquid hydrocarbons of any type have been found in those sediments.

Carbohydrates, Proteins, Fats

Regardless of type of plant or animal source, there is also the question of what part or parts of those organisms could possibly have been the source material for petroleum. Carbohydrates are generally thought not to have been an important source. That is mainly because they are soluble in water. Geologic findings suggest that most petroleum was formed related to a marine environment. In the biologic origin theory carbohydrates would be expected to have dissolved prior to production of oil. Carbohydrates are essentially absent in modern marine sediments. They are not found even in traces in petroleum or oil field water.

Proteins have also been dismissed as significant sources for petroleum. Proteins would probably have decayed as a result of autolysis before conversion to petroleum could have occurred.

Organic theory proponents have postulated that fats are the principal source of material for petroleum. In additiion to all of the evidence presented above against any organic biologic origin of petroleum, there is other strong evidence against such origin particularly applicable to fats.

Attempts to decompose fatty acids to methane have failed. Analyses of numbers of carbon atoms in fats and in petroleum have yielded evidence that seriously damage any remaining credibility of the theory. The fatty acids and waxes of living organisms have an even number of carbon atoms. If those fatty acids were converted to paraffins and other hydrocarbons, one would expect almost all of the resulting compounds to have either an even or an odd number of carbon atoms depending upon the chemical process that was involved in the conversion. There are claims that there are more hydrocarbons composed of odd numbers of carbon atoms in paraffins of younger crudes. One must take all of the hydrocarbons into account, and they are much easier to analyze when dealing with crudes high in paraffin content, which requires comparisons of older crudes since paraffin content increases with age. A careful analysis of the reported findings reveals that there is no real consistent significant preponderance of either odd or even numbered carbon atoms in paraffins and other hydrocarbons.

Fats found naturally occurring are mostly composed of fatty acids having 16 and 18 carbons. In contrast, a significant quantity of petroleum is composed of molecules having fewer than 10 carbons. Further, the heaviest of waxes found in living organisms have 34 carbons in a molecule while molecules having 50 and 70 carbon atoms have been found in petroleum.

A large percentage of fatty acids of living organisms are unsaturated. That means that there are double bonds between carbons where hydrogen atoms could be added. Unsaturated fatty acids are not found in petroleum or in water with petroleum. Other fatty acids are found in those liquids. There are also no unsaturated straight chain paraffin hydrocarbons in petroleum. It is clear that common fatty acids do not fit into the organic theory of petroleum origin either as to type of carbon bonds or numbers of carbon atoms in their molecules.

It has been generally recognized that the occurrence of the large amounts of cycloparaffins in petroleum cannot be satisfactorily explained by the organic biologic theory of petroleum formation. Cyclopentanes and cyclohexanes, common in petroleum, are so rare in nature that it has been considered inconceivable that they can be accounted for in petroleum by any kind of degradation of material found in living organisms.

From the foregoing, it is clear that the findings of porphyrins in petroleum and optical activity of petroleum are not evidence in favor of the organic theory. Evidence that petroleum could have come from various types of organisms is extremely defective. The main support that the theory has are basically similar gross appearances of fats, coal, and petroleum and lack of a convincing inorganic abiotic theory of petroleum origin. Geologic and chemical findings overwhelmingly refute the theory that petroleum was produced from formerly living organisms. A sound scientific theory of abiotic inorganic origin of petroleum follows.

IX. INORGANIC ORIGIN OF PETROLEUM

It is proposed that petroleum results from gases expelled from the substance of the Earth. Heat in the inner Earth is a recognized observation most obvious with volcanic activity. The heat is probably mainly due to pressure, contributed to by chemical and nuclear reactions. Gravitation causes the substance of the Earth, including water, to exert great pressure toward the center of the Earth, increasing with depth in the Earth. Pressure produces heat. The average increase in temperature of the ground is one degree Celsius per 100 feet of depth. It is postulated that heat causes gases to form from hot molten magma which if unimpeded would bubble to the surface. The crust of the Earth, and particularly where it is overlaid with deep bodies of water, obstructs the free release of those gases into the atmosphere.

Water readily transmits pressure, so at depths of oceans, the pressure exerted on the crust of the Earth is tremendous. The pressure of the air at sea level, one atmosphere of pressure, is 15 pounds per square inch. Five miles below the surface of the ocean the pressure is 800 atmospheres. It is proposed in this new theory that basins or depressions in the Earth's surface, and particularly where covered by water, cause the bubbles of gas from the deep magma to move along the generally upward sloping paths to the edges of those basins. At the edges of those basins if the crust is weak or the rock is too dense to accommodate the gases, the bubbles of gas are released as volcanoes with or without accompanying magma. If the bubbles reach porous matter such as sand or gravel which is capped by dense rock, the gases will not bubble into the atmosphere but will react and join with each other polymerizing to form the complex hydrocarbons of petroleum. With the oceans covering approximately 70% of the surface of the Earth, they are responsible for most of the concentrations of gases to form petroleum or to be released through volcanoes or other vents.

The foregoing theory explains why significant petroleum fields are not found where there is volcanic activity. It accounts for the small hydrocarbons, particularly methane, found commonly in volcanic gases. It explains why volcanoes and oil are found in geologically similar and related situations.

Much confusion has resulted from so-called "mud volcanoes." Apparently rare, they are spectacular in the Baku region of the U.S.S.R. Hydrocarbon gases and liquids burst through the surface of the Earth in those "mud volcanoes." Occasionally the gases catch fire and burn, but generally they are cold. Those eruptions occur *after the conversion* of the magma *gases to petroleum.* The gases of mud volcanoes are not under much pressure. It would be comparable at the greatest to the pressure of "gusher" oil wells. That is negligible compared to the pressure of gases in real volcanoes. The gases that are released in mud volcanoes do not come directly off the hot magma as the gases do that form real volcanoes. There is no accompanying flow of magma. Because they are hydrocarbon "volcanoes," they have caused confusion of the correct relationship between petroleum and real volcanoes. The gases of real volcanoes in the theory are from the hot magma before they could be converted to petroleum. The gases of mud volcanoes are released after conversion to petroleum and are comparable to other natural seepages of petroleum to the surface of the earth.

If large bodies of water did not form large basins, the surface of the Earth would probably be covered nearly uniformily with volcanoes and craters from release of gases from the inner earth. In this theory, the presence on Earth of large deep bodies of water is the main reason for the concentration of the gases to form petroleum. Also, the large land basins which have a concentrating effect similar to ocean basins probably would not have been formed but for bodies of water. Water has also been important in producing the porous traps of sand and gravel which were produced mainly from water erosion of rocks.

The planet Venus, which has no bodies of water, has an atmosphere mainly of carbon dioxide. The probable similar carbon content of Earth has not been released into the atmosphere. According to the presented theory, the primary reason it has not been released is that it has been trapped to be released occasionally through volcanoes and to be converted to petroleum in other locations.

That is the basic theory. No ideas are probably entirely new or exclusive to an individual. Prior to 1900 the generally accepted view was the petroleum (Latin, *petra oleum* — "rock oil") was inorganic, "mineral," in origin. In 1804, Alexander von Humboldt hypothesized that petroleum is the "product of a distillation from great depth and issues from the primitive rocks beneath which the forces of all volcanic action lie."* Volcanoes were considered to be the origin of petroleum by Lenz, a Russian, in 1831, and by the French geologist Virlet d'Aoust in 1834. Eugene Coste, in 1903, elaborated on the same idea of relating hydrocarbon emissions from volcanoes to petroleum and its gases. Berthelot, in

**Dott, p. 15*

1866, postulated that petroleum could be the result of steam acting upon carbides producing gas within the Earth. Strabo, a Greek who died in 23 A.D., reported that ancients considered petroleum to be of volcanic origin. Gases have been recognized by others as being the primary cause of volcanic activity.

As for volcanic activity on ocean floors, it is apparently primarily effusive, lava flow without much gaseous element. The fluid dynamic relationships between liquid magma and water separated by a crust are poorly appreciated. Deep ocean bottoms are surrounded by liquid magma at higher level held up there by water pressure. The force of the water is to seek the center of Earth due to gravity. The force of the magma, due to gravity, is to flatten its own surface. Cracks in the crust in an ocean basin may result from either of those forces with variables such as increasing or decreasing depth of water. With increased water depth the downward pressure on the crust would increase, the ocean would deepen further, a crack in the crust would result, it would widen, and the water pressure would thus cause water to displace magma. Magma would usually be expected to be pushed through the crust somewhere else. With decrease in the depth of the water due to decrease in amount or its being spread over a greater surface, a crack would result from magma pressure from below and it could then flow through that crack onto the ocean floor. Any release of gas or magma flow changes the balance of the force of water and the force of magma and can cause cracks in the Earth's crust, their widening, or other adjustments.

A relationship between the seas and volcanoes was recognized as early as the sixteenth century by the Italian philosopher Bruno who noted that most volcanoes were near oceans. Prior to the theory presented herein, there has been no satisfactory explanation for the relationship between oceans and the cause of volcanoes. It has been generally accepted that the oceans primordially came from water expelled from volcanoes. Tectonic movements in the crust can have nothing to do with the massive accumulations of gases near oceans that erupt as volcanoes. As for sites of cracks, a crack in a rigid structure requires bending or a crack somewhere else, depending upon the degree of rigidity. One crack in the ocean floor caused by water pressure may require other cracks. The pressure of water on the ocean floor probably acts as a splint and prevents some such compensatory cracking which then occurs near oceans.

The main reasons that the volcanic theory of petroleum origin fell out of favor were the common observations that petroleum is not found in significant amounts where there is volcanic activity, it is not found in significant amounts in rocks of volcanic origin, and large amounts of petroleum are found where there is no volcanic activity. Those objections are met and accounted for in this new theory. Both volcanoes and petroleum are from the same gaseous source, but where the gases are released in volcanoes an appreciable amount of petroleum cannot be formed.

There is considerable evidence that hydrocarbons can come from the rocks which comprise the elementary substance of the earth. Carbon and hydrocarbons are found in igneus rocks, rocks of volcanic origin. As early as 1877, oil and solid paraffins were found in basaltic lava near Etna. Similar findings were reported from Java in 1909. Even complex hydrocarbons including waxes have been found in meteorites. The abundance of methane in the atmosphere of Jupiter and Saturn may also be considered as evidence that the substances of the planets can produce hydrocarbons.

Although not common, petroleum and natural gas have been found in commercial quantities in volcanic primordial rocks, igneus and metamorphic basement rocks. It has been claimed that methane, ethane, and propane can be found almost anywhere deep wells are drilled, and therefore basement igneus rocks have been claimed to be the probable source.

Mendeleeff demonstrated in 1877 that dilute acids or hot water acting on iron and maganese carbides resulted in formation of petroleum-like hydrocarbons. Acetylene, C_2H_2, from such reactions has been treated to produce larger hydrocarbons. Steam acting on calcium carbide yields acetylene and acting on aluminum carbide yields methane. Liquid hydrocarbons have been produced from carbides of uranium and rare earth elements.

From the foregoing, relationships between volcanoes and oceans and between volcanoes and petroleum have been suspected and postulated in various ways since ancient times. The presented theory connects all three, volcanoes, oceans, and petroleum, in an original, logical way. It postulates releases of gases from within the hot depth of the earth. Because of the tremendous pressures exerted by oceans on the underlying crust of the earth, most of the gas formed under the oceans migrates to the edge of the oceans. If the gas reaches a suitable porous reservoir, it may be stored there as gas; or, the gaseous molecules, affected by the heat, pressure, and clay catalysts present, join together by polymerization to form petroleum. If the gas does not reach clay, sands, gravels, and an adequate reservoir, and it reaches a place where the crust of the earth is weak from cracks of faults, fissures, tilting, or other features with thin spots in the crust, the gas is released through volcanic activity.

An examination of the locations of volcanoes and petroleum accumulations in relationship to oceans is consistent with this theory. As noted, the ancients associated volcanic activity with seas. Seven per cent of the presently known active volcanoes are about the Mediterranean Ocean. That makes up the bulk of the volcanoes of Europe. Another 13% of the known active volcanoes are located on islands in the Atlantic Ocean. The "ring of fire" surrounding the largest body of water, the Pacific Ocean, is well recognized. It accounts for 62% of the known active volcanoes on Earth and another 3% occur on islands of the central Pacific. The Indian Ocean accounts for another 1%. Almost all volcanoes in the world are within about 100 miles of oceans.

Where there is volcanic activity as a part of the Pacific "ring of fire" in Oregon, Washington, and Northern California, there is essentially no petroleum. On the other hand, in Southern California where there are no volcanoes, there are tremendous accumulations of petroleum. The great petroleum accumulations about the Carribean, including Texas, Mexico, and in the continental shelf off their shores, are in areas where there is no volcanic activity. There is no volcanic activity in the region where petroleum has been found in the North Sea. There is no significant petroleum near Etna or Vesuvius; but it is plentiful in the Middle East, especially in relationship with the Indian Ocean where there is essentially no volcanic activity. Japan has at least 33 active volcanic vents and no appreciable petroleum. Hawaii is primarily volcanic in origin and has no commercial deposits of petroleum.

The formation of the islands of the central Pacific are consistent with the theory. The Pacific is so huge it is reasonable to anticipate the gaseous and magmatic eruptions in its center producing islands and basaltic mountains upon which coral have been able to form other islands. It was simply too far for the gases to migrate to reach the edge of the ocean without accumulating enough gas volume and under enough pressure to break through the pressure of the ocean on the crust of the earth.

Extremely important, this theory is consistent with and explains the scattered distribution of the continents. One large land mass originating about the North Pole and another about the South Pole, as often postulated without real evidence, is inconsistent with this new theory of the cause of volcanoes. Gases accumulated at sites under the oceans more or less equidistant from each other and erupted with flows of magma to form the large land masses at approximately their present locations. It may be noted that there is no Coriolis effect apparent in the distribution and shape of continents. There is no evidence of land nearer the equator having moved farther than that more distant. Continental drift has been minimal and at the most has caused only slight wrinkling at some edges.

As might be expected from the theory, gases formed beneath land masses behave similarly to those under oceans except for the decreased ability of the crust to contain them without the overlying water. Surrounding the land basin of the center of the United States, oil accumulations are found in Colorado, Wyoming, and in Canada in areas where there is essentially no volcanic activity. The great plain of the Snake River which slopes upward from west to east approximately 300 miles across the State of Idaho is topped in the east by the volcanic area known as the "Craters of the Moon." With no large bodies of water overlying it, multiple flows of lava have occurred throughout the plain. No commercial quantities of petroleum have been found in the region.

The amount of gas blown off in volcanoes is consistent with what could have otherwise been large pools of petroleum. In the 1906 eruption of Vesuvius the weight of the gases was estimated to be many times greater than the total weight of ash and lava.

The nature of gases vented from the earth through fumaroles and volcanoes is consistent with the theory. The odor of burning bitumin has been reported detected from volcanoes. Studies of gases from volcanoes and the related fumaroles are limited, but they reveal that the gases vary somewhat in chemical nature from different volcanoes and at different temperatures from the same fumarole or volcano. Certain elements and chemical molecules are common in those gases. Water and carbon dioxide are invariably abundant. With the tremendous heat and oxygen available from oxygen compounds in magma, carbon dioxide and water are not unexpected. Also, gases from magma have been observed burning. Sulfur in the form of hydrogen sulfide and sulfates are also common. Ammonia is frequently present, and in some fumaroles ammonium chloride has accumulated in commercial quantities. Other gases found include methane, nitrogen, helium, argon, hydrochloric acid, chlorine, and many others.

It has been recognized that most of the more volatile compounds were released from the magma before it cooled, but bubbles in volcanic rocks have been found to contain water, carbon dioxide, carbon monoxide, methane, and nitrogen.

Pressures of gases of volcanoes can be extreme. The eruption of Tomboro on Sumatra in 1815 is estimated to have blown up 50 cubic miles of earth, equivalent to 185 mountains the size of Vesuvius. About 5 cubic miles of material was blown up in the Krakatoa eruption in 1883. The recent explosions of Mount St. Helens in the State of Washington attested to the tremendous pressures of the gases.

Temperatures of fumaroles have been recorded to be as high as 900 degrees Celsius. Magma has been found to have temperatures from 800 degress to 1,500 degrees Celsius.

It has been clearly shown that the chemicals necessary for the production of petroleum are present in the substance of the earth and in sufficient quantities to account for the huge amount of petroleum known to exist. The temperature within the Earth is sufficient to release those chemicals as gases. Evidence is abundant that gases are in fact released from the hot interior of Earth. The gases known to be released include hydrocarbons, all of the elements necessary to form petroleum, and gases commonly found with petroleum.

The postulate is reasonable that some gases formed within the substance of Earth are not released into the atmosphere, but are trapped in clays, sands, and gravels capped by hard impervious rock. The postulate is logical that under similar circumstances that polymerization of hydrocarbon gases has occurred and produced petroleum. Necessary heat, transient pressure, and catalysts were present in the postulated circumstances to have caused such polymerizations and produced petroleum.

Geographic locations and geologic relationships between the water basins and the land basins of the planet, petroleum fields, and volcanoes are consistent with the theory.

The next key concept or postulate concerns the chemical changes to form petroleum and changes after it is formed. As the hot bubbles of gas from the magma contact porous sands and clays three significant processes occur. First, the gas starts to cool from the 1,500 degrees Celsius of the magma. Second, the pressure against the gaseous molecules starts to decrease. Third, as the bubbles of gas reach lower temperatures in contact with sand, clay, and mineral catalysts they contain, the molecules in those bubbles react with each other and form all of the diverse chemical compounds found in petroleum. The process of cooling allows for chemical reactions to take place at the optimum temperature for each of those reactions. The cooling continues until the petroleum reaches the temperature of the ground at that depth. It then migrates to a reservoir that is usually cooler.

The high temperature is a key factor in forming petroleum and to be presented further later, in forming the chemicals of life. It is elementary in chemistry that chemical reactions proceed in both directions. The final amounts of starting materials and new products depend upon conditions such as catalysts to enhance a reaction, heat, and pressure. When the chemical reaction reaches a point where the reaction proceeds at an equal rate in both directions it is said to be in equilibrium. The following is a simple chemical equation to illustrate that principle:

$$CH_4 + CH_4 \rightleftarrows C_2H_6 + H_2.$$

The arrows indicate that the reaction proceeds in both directions. Starting with two molecules of methane, CH_4, one molecule of ethane, C_2H_6, and one molecule of hydrogen, H_2, can be obtained. Or in reverse, starting with one molecule of ethane and one molecule of hydrogen, two molecules of methane can be obtained. Such a reaction proceeds in both directions and the final result depends in part upon the temperature. By way of a simple hypothetical example for demonstration purposes only, starting with two thousand molecules of methane at a temperature of 400 degrees Celsius, the final resulting equilibrium might be something like 1000 molecules of methane remaining and 500 molecules of ethane and 500 molecules of hydrogen produced. That final equilibrium of the chemical reaction is those amounts of the different gases.

In general the equilibrium studies of petroleum have been a source of confusion. Isomers of hydrocarbons are compounds which have the same number of carbons and hydrogen but they are arranged differently. The amounts of different isomers of hydrocarbons containing 8 and 9 carbons is similar for various crude petroleums and corresponds to an equilibrium at 455 degrees Celsius. In addition to conflicting with the organic biologic theory of origin of petroleum, the findings do not appear to fit with the temperature of over 500 degrees required to produce aromatic hydrocarbons. Important but apparently adding to confusion, equilibriums of isomers of hexanes (6 carbons) and heptanes (7 carbons) correspond with temperatures above 200 degrees Celsius in some petroleum samples and with temperatures below 200 degrees in others. An even more important observation is that all of the complex hydrocarbon molecules in petroleum are unstable, not in equilibrium, at the temperatures and pressures in which they are found.

In this theory, petroleum was formed under changing conditions of temperature and pressure that did not allow all of the chemical reactions to reach equilibrium. Compounds requiring high temperatures formed, the temperature dropped and different chemical reactions occurred. The final products were unstable, or not in equilibrium, at lower temperatures; so they have proceeded toward stable compounds. Examples of the stable molecules for petroleum are carbon, C, methane, CH_4, and ethane, C_2H_6.

It appears clear that chemical molecules in petroleum probably started to deteriorate from complex hydrocarbons including complex ring structures to become paraffins, including methane and ethane, as soon as the temperature dropped after they were formed. The longer the petroleum is stored at temperatures lower than that at which it was formed the more such deterioration can be expected. This correlates with the common observation that older more ancient crudes usually have a higher content of paraffin compounds while younger crudes are richer in compounds with complex ring structures.

It is submitted that the proposed theory of production of petroleum from gases released from the hot molten substance of the interior of Earth has been shown to be the best explanation of petroleum formation. It is logical, based upon sound scientific observations, experiments, and established principles, and is consistent with and encompasses all relevant observations. With this foundation, the theory of origins and creations of life in all of its varied forms from petroleum will be presented.

X. THE CHEMICALS OF LIFE

The essence of life is unknown. Regardless of lack of that specific information, there are observations from which conclusions may reasonably be drawn concerning the probable origins of living organisms. Certain chemicals are intimately connected to life. The most important ones seem to be the nucleic acids. Under conditions within cells and the chemicals there, the nucleic acids direct the process of protein synthesis from amino acids. The nucleic acids in germ cells such as ova and sperm contain the "genetic code" and scientific experiments have revealed that they probably direct the process of anatomic development and determine the physiologic capabilities of all body parts of the most complicated living organisms, both plants and animals.

The genetic code determines the characteristics by which organisms are classified as to class, order, family, genera, species, and subspecies or varieties. It determines whether the organism becomes a plant, fish, mouse, or human. It determines whether reproduction is accomplished by just simple cell division as bacteria do, or whether reproduction requires the joining of ova and sperm from different animals. In the human it is the genetic code that directs the total reproductive process including conception, implantation in the uterus, gestation including development of the embryo, birth, and nutrition following birth.

All of the foregoing direction and control of organisms are contained in single germ cells such as ova and sperm. Single cells can reproduce more similar single cells or they can produce the most complex living organisms depending upon the genetic code and whether they are body cells or sex-germ cells.

In the theory presented here it is postulated that differences in living organisms resulted from differences at the time of the origin of the germ cells from which they were formed. Considering the complexity of chemical reactions required for life to originate, it would require only minor differences to account for the variety of living organisms of the past and present. Minor differences in temperature and in chemicals involved in forming life including those in petroleum, gases, water, or clay could account for the differences of all living organisms.

In this theory conditions were ideal or nearly ideal to form living organisms in many places at many times, and those conditions included an abundant supply of materials and in proper concentrations. The sites were the many places where freshly formed hot petroleum and petroleum gases reached the surface of the earth. At those places when the conditions were right and the simplest form of living organism could originate, it is postulated that many of those and similar units of life were formed. Some of those simplest units of life stuck together to their mutual advantage and the result was cells. Some of those single-celled living organisms were germ cells of many types. Nutrients were plentiful and the complex forms of life which developed from those germ cells thrived.

By contrast, the theory of origin of life generally accepted by scientists postulates that life started in meager circumstances where source materials were in dilute concentrations. It postulates that the source materials were gases in the atmosphere. Only the simplest form of life is postulated as having started at the site of origin. Encompassing the theory of evolution, it then postulates that the various forms of life developed from that simplest organism by changes in succeeding generations.

Closer examination of the chemicals involved, chromosomes, and living organisms on different continents and islands will be shown to support the theory of multiple creations and to discredit the theory of evolution.

The most important chemicals of life are apparently deoxyribonucleic acid (DNA) and ribonucleic acid (RNA). Ribonucleic acid (RNA) is smaller than DNA and instead of containing the sugar deoxyribose it contains ribose. A large DNA molecule is about 100 times larger than a large RNA molecule. One of the chemical bases of RNA is also slightly different from one of the bases of DNA. DNA and RNA are large polymers resulting from the joining together of a few relatively simple chemicals. With rare exceptions, DNA is composed of only four chemicals which are joined together in a variety of sequences. The differences in sequences of those four chemicals in the resulting strand or chain is the genetic code. The length of the strand may consist of as many as 10,000,000 of those units. If stretched out, it would be about a yard, or about one meter, in length. RNA is also composed of only four chemicals similar to those of DNA joined together.

Proteins which accompany the DNA and RNA are also usually strands composed of comparatively simple chemicals joined together. They are composed of up to 20 amino acids, most of which are relatively easily produced from such simple compounds as ammonia, methane, and water. The production of proteins from amino acids in cells is accomplished under the direction of DNA. Which amino acids are used to make a particular protein and their sequences depends upon the combinations in sequences of the four chemicals in the DNA strand. In complex cells RNA actually accomplishes the protein production with the directions or code supplied by the DNA.

A small virus may consist of only RNA when it is inside a cell. Bacteria may consist primarily of DNA. More complex cells contain DNA and RNA in various locations where they perform various specialized functions. Those more complex cells may be thought of simplistically as conglomerates of units of life such as multiples of DNA and RNA surrounded by a membrane. It is postulated here that they were formed initially in that manner. Direction for construction and the construction of membranes which separate parts of the cell from other parts and membranes which surround the cell are probably accomplished by the DNA and RNA. The membranes generally consist of proteins sandwiched between layers of phosphoilipids, fat-like compounds. The first templates or structures upon which membranes were formed could have been as simple as a film of oil to separate the cell or its parts from the air or water environment. It could have been the surface of a colloid gel or "coacervate" as proposed by Oparin, the surface of a form resulting from adsorption of hydrocarbons onto a particle of sand as proposed by Bernal, an oily film as proposed by Haldane, or the surface of proteinoid material gathered in "microspheres" as postulated by Fox. A variety of means and mechanisms may have been involved in organizing the variety of forms of life.

DNA is the most complex and apparently the most important of the chemicals of life. The four chemicals of which it is constructed are called nucleotides. Each nucleotide in turn is composed of three chemicals, only one of which varies. The variable is the "base." The bases are purines and pyrimidines. The purine and pyrmidine bases of the nucleotides in DNA are adenine, guanine, cystosine and thymine. They are hydrogen-carbon ring chemical compounds. The basic structure of the group of the group of pyrimidine bases is depicted in *Figure 8.* In the depictions of the chemical compounds in the figures, C stands for a carbon atom, H for a hydrogen atom, O for an oxygen atom, and N stands for a nitrogen atom. The single and double lines refer to bonds or the electrons shared between two atoms in their joining together. The understanding of the nature of that bonding is not necessary to understand the principles of this presentation.

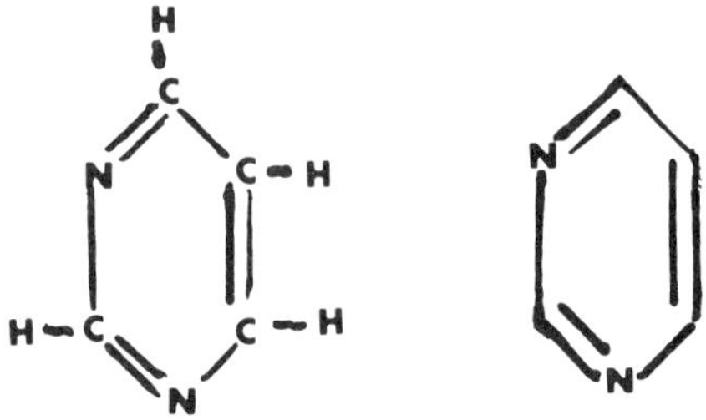

Figure 8. The basic structure of the pyrimidine group of bases. On the right the same structure is depicted without carbons and hydrogens designated.

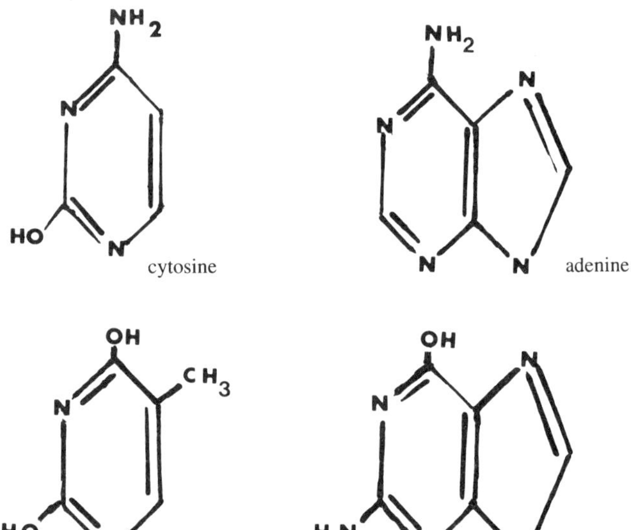

Figure 9. Pyrimidine bases, cytosine and thymine, and purine bases, adenine and guanine.

The four bases in DNA, cytosine, thymine, adenine, and guanine, are depicted in *Figure 9.* From their chemical structures it is apparent that the bases are all similar and probably they were originally formed in the same or similar chemical reactions.

To form a nucleotide, each base is attached to a sugar and to a phosphate group. The sugar that is attached to each of the bases in DNA is also a carbon-hydrogen ring compound named deoxyribose. It is depicted in *Figure 10.* Deoxyribose added by a chemical reaction to adenine forms the nucleoside named deoxyadenosine. It is depicted in *Figure 11.* To complete the nucleotide, a phosphate group is added to deoxyadenosine. The completed nucleotide is depicted in *Figure 12.*

The other three nucleotides are constructed in the same way, the only difference being that a different purine or pyrimidine base is attached to the sugar. As noted above, the four nucleotides are joined together in a long strand by joining their phosphate groups together. Then two such strands are paired and joined together side by side and arranged in the form of a long double helix, the shape of a spiral, much like a spiral stair with a rail on each side.

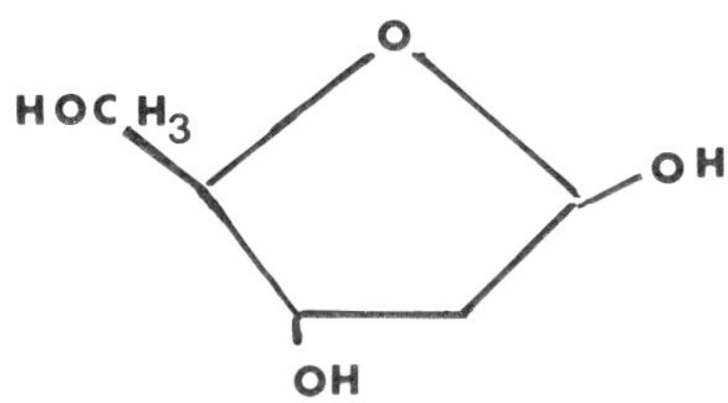

Figure 10. Deoxyribose, the sugar of deoxyribonucleic acid (DNA)

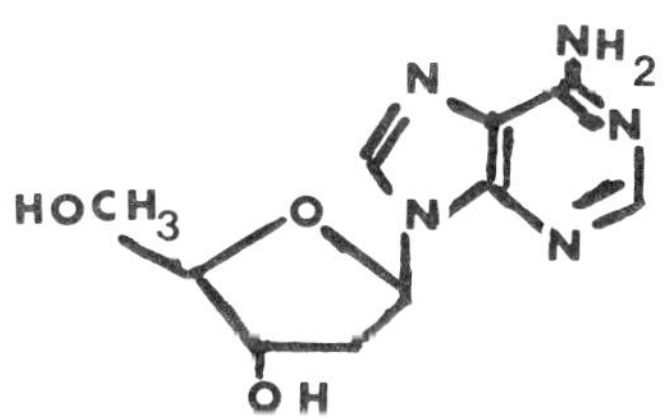

Figure 11. Deoxyadenosine, a nucleoside of DNA resulting from deoxyribose and the base adenine.

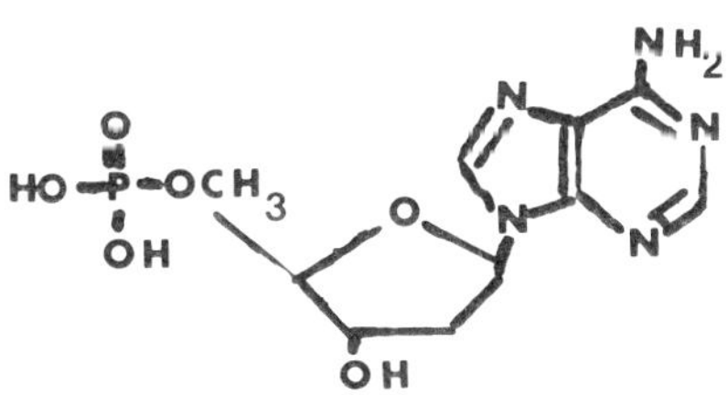

Figure 12. The nucleotide deoxyadenylic acid.

Once the linkage of four nucleotides started in the primordial environment, it seems reasonable that it could continue to build various necessary lengths. There are certain sequences of nucleotides that do not serve a function in protein synthesis and terminate the strand. It is postulated that formation of DNA from petroleum and other material will probably prove to be rather simple when it is finally understood. However, its formation requires several chemical reactions and they are of such a degree of complexity that variations in the combinations of the nucleotides would undoubtedly result.

Protein is formed by joining together in various sequences twenty different amino acids. It has been shown that most of those amino acids can be produced from simple chemical reactions between methane, ammonia, and water. Enzymes that enhance the formation of DNA and various metabolic functions in living organisms are specialized proteins.

Several groups of scientists have performed many experiments directed to producing the important chemicals of life including amino acids, proteins, sugars, nucleotides, DNA and RNA. Many of those experiments have been performed with limitations imposed in accord with the imagined conditions at the time life first appeared on Earth. The temperature of the postulated environment was less than the boiling point of water which is 100 degrees Celsius. The postulated atmosphere consisted mainly of methane, ammonia, nitrogen, and water. There were oceans and there were land masses probably without mountains except for low volcanoes. The land was mainly rock of volcanic origin with some sands and clays which were the result of erosion of the rock.

Under experimental conditions simulating that postulated primordial environment more than 25 amino acids have been produced.

Porphyrins have also been synthesized from methane, ammonia, and water. That not only proved that the base for hemoglobin and chlorophyll could be produced under primordial conditions, it also proved that the presence of porphyrins in petroleum can be explained by a simple chemical process. It proved that porphyrins in petroleum are not evidence of origin of petroleum from formerly living organisms. The same substances of which porphyrins are produced in the laboratory are present in gases from volcanoes and fumaroles.

Sugars including ribose and glucose have been synthesized under the restricted conditions.

Adenine has been synthesized from ammonia and cyanide. Cyanide has been observed in tails of comets, so it has been allowed in experiments with postulated primordial conditions. Syntheses of adenosine and cytidylic acid have not been accomplished in such conditions.

Purines heated with ribose have yielded nucleosides. No pyrimidine nucleosides have been similary produced. A major problem has been the restriction to alkaline solutions.

Adding phosphate has proved difficult. The common salts of phosphates do not readily add onto the other chemicals. If phosphoric acid or acid phosphates are used in dry mixtures of nucleosides (base plus sugar), phosphates are added to them to produce nucleotides and with some joining together or polymerization of the resulting nucleotides.

Proteins are polypeptides which require joining amino acids together. Protein-like compounds have been produced by heating dry mixtures of amino acids to 180 degrees Celsius, but not protein. Even the protein-like compounds have not been produced in the postulated primordial conditions.

Fats are compounds of fatty acids joined to glycerol. No fatty acids have been synthesized in the postulated primordial conditions. Also, no glycerol has been produced under those conditions. Industrially it is made from propylene, an unsaturated three-carbon hydrocarbon.

An evaluation of the varieties of the theory of origin of life from an atmosphere of methane, ammonia, nitrogen, water, and perhaps carbon dioxide and cyanide with a resulting slick on the surface of the ocean, foam on the water, and congealed amorphorous matter is in order. The most brilliant chemists in the most sophisticated chemical laboratories in the world have been unable to produce essential chemical compounds required for life in conditions simulating their postulated primordial Earth and their postulated origins of life. Recognizing an apparent impasse, it has been postulated that chemical reactions which cannot be accomplished in modern chemical laboratories were in some mysterious manner accomplished by "chemical evolution" in nature over a period of more than one billion years. That effectively removes that theory of origin of life from scientific scrutiny. It removes it from the usual disciplines of science.

The most lenient evaluation possible is that the theory is reduced to the status of speculation and remains tentative depending upon any supporting evidence that might be produced and is presently lacking. A reasonable objective evaluation would be that the theory at this point has been proven to be wrong.

This new theory retains all of the chemical materials and adds more to the aforementioned theories; and it adds new conditions. Ammonia and methane can bubble into pools of water. The water can be fresh or salty. Temperatures up to 1,500 degrees Celsius and anywhere below are available. Many of the esential chemical reactions would have been performed in the production of the petroleum before it reached the surface of the earth. Porphyrins and fatty acids would have already been produced. Complex ring hydrocarbons already formed could be altered to form various similar compounds found in living organisms. The hot petroleum could have been sprayed or bubbled onto rocks or dry clays or into water, or there could have been combinations of those events. The problems of adding phosphorus and of joining phosphates together to form the long DNA strand might be as simple as the seventeenth century alchemy separation of pure elemental phosphorus by heating urine mixed with clay and charcoal. Hot petroleum spilled onto rocks rich in phosphates or onto phosphate salts might yield elemental white phosphorus as in the original experiment. Elemental phosphorus forms interesting compounds with hydrocarbons as well as forming acid phosphates. The chemistry of phosphorus is difficult and complex. A great deal of experimentation may be required. When the correct sequence of events is learned it will probably be simple and commonly repeated in nature and easily repeated in the laboratory.

It is postulated that when the conditions were such that DNA, RNA, and protein could have been formed, it would have required conditions that caused them to be produced in abundance and variety. With those units of life in abundance and variety, the formation of germ cells and ova probably did not require much more. It is postulated that all varieties of plants and animals, with the possible exception of some subspecies, originated in that manner.

As explained above, gases that were vented through volcanoes or became petroleum were the same types of gases that migrated into different types of geological formations. Petroleum seepages and "mud volcanoes" are eruptions of gases and petroleum after petroleum has been formed. In their mature forms, those are three distinct phenomena, volcanoes, petroleum production, and seepage of petroleum and its gases or eruption of them in mud volcanoes. Gases from such mature volcanoes may contain methane in small amounts, but larger hydrocarbons have not been reported. The gases from the St. Helens explosion in 1980 were apparently not as hot as magma, 800 to 1500 degrees Celsius, or the timber it felled would have burned. Therefore, it cannot be assumed that petroleum would necessarily be burned if formed in association with a volcano.

When Earth was young and sediments were thin, the distinction between the production of volcanoes, petroleum, and seepages or mud volcanoes was probably not as sharply defined as now. As the gases moved from magma into a sediment and formed petroleum and hydrocarbon gases, the sediments probably soon could not contain the gases under mounting pressure. The resulting eruption then included hot volcanic-type gases, petroleum, and magma. Similarly when a volcano erupts for the first time, its gases may have passed through enough sediment to have converted some of the gases to petroleum. The volcanic spewing and spattering of hydrocarbons over wide areas has probably resulted, according to this theory, in production of plants and animals that have short incubations or gestations. Other organisms, as will be further explained, probably required more stable conditions that were attained from petroleum seepages.

The early explosive eruptions of volcanoes containing petroleum probably produced the many species of multicellular complex organisms which occurred so early at the base of the Cambrian period. Only the precambrian layer is below it.

It has been claimed that changes of types of living organisms have corresponded to starting of new sediments or geologic periods. Species persisted through a stratum and then seemed to have been suddenly replaced. The change of geologic periods has been connected to tectonic events. Movements of continents, volcanic eruptions, mountain upheavals, possible changes in the rotation of the Earth, and similar events could be expected to cause cracks in the crust of the earth releasing fresh, hot petroleum in seepages and through new volcanoes. A new crop of plants and animals could be expected at those times in the new theory. At the end of the Permian period, 225 million years ago, apparently most of life was wiped out. Different plants and animals followed. The change in the geological period was probably associated with more fresh leaks of petroleum as well as new volcanoes.

The petroleum that has erupted with volcanoes has probably not had the same constituents as petroleum that was able to stay underground long enough to complete the chemical reactions to produce the more complex molecules. In other words, it may have erupted to the surface before it was the complete petroleum product with the entire spectrum of chemicals. With better containment by more overlying sediments, more complete petroleum was produced including its most complex hydrocarbons. Seepages of the more complete petroleum and its gases became less explosive and lasted longer. As layers of sediment have become thicker and petroleum has been trapped by overlying impervious rocks, the seepages of petroleum have been mainly the result of faults in the crust of the earth, movements of the crust, and earthquakes. Later seepages along fault lines were probably in natural depressions and formed more permanent, stable pools. Slower seepages probably contributed to more stability in the temperatures of the pools.

Those changes are probably largely responsible for some of the general differences between the earliest living organisms and later arrivals. Examples of the results of the various types of seepages may still be observed. Young volcanic islands probably present conditions similar to the primordial Earth in many respects. Well formed petroleum seepages are not found on the Hawaiian Islands. It is postulated that most of the plants and animals native to those islands originated there from explosive eruptions containing petroleum similar to those of Earth in its earliest geologic periods. The animals of the Hawaiian Islands consist of land snails, insects, and birds. The Galapagos Islands are similarly of recent volcanic origin. The highest forms of native life there consist of iguanas, tortoises, a snake, and birds. Australia presents conditions intermediate between other continents and the primordial state of young volcanic islands. It has a low profile, with its highest mountain being only 7,316 feet high (2,230 meters). It has had no great movements in its crust in recent geologic periods. It experienced none of the geologic "revolutions" that formed the Himalayas in Asia, the Alps in Europe, the Rocky Mountains in North America, and the Andes in South America. As a result, it probably missed out on having the most recent seepages of petroleum. Those were the seepages that were probably necessary for the originations of the placental mammals. Its relatively primitive flora and fauna are natural consequences.

Seepages of petroleum to the surface of the Earth are common. The famous Pitch Lake in Trinidad is about 2,000 feet in diameter and is as deep as 135 feet. Bermudez Pitch Lake in Venezuela covers more than 1,000 acres with an average depth of 5 feet. Most of it is covered with vegetation. There are numerous seepages in Southern California from Los Angeles to Coalinga, a distance of 140 miles. The most famous, the La Brea Tar Pits (named for a similar occurrence in Peru), are in Los Angeles. Seepages are common in the Middle East, including in Iraq, Iran, and Saudi Arabia, and in the Baku area of the U.S.S.R. At least 5 oil fields in Iran have overlying seepages. The Dead Sea was formerly named Lake Asphaltes because of petroleum seepages there. The most famous seepage in ancient times was at Hit on the Euphrates River in Iraq. The tar sands on the Athabasca River in Alberta, Canada, are outcroppings which stem from sandstones impregnated with petroleum in an area of thousands of square miles. Numerous small seepages have occurred in Burma, China, at Pechebronn in Alsace, Grosni in the U.S.S.R., Wietze in Germany, in Switzerland, and around the Caucasus Mountains. Consistent with the new theory herein, it has been noted that marine seepages are rare. They have reportedly occurred offshore from Santa Barbara, California, and from Trinidad, and in the Gulf of Mexico and in the Caspian Sea.

Originations of life are not abundantly evident at those seepages because they are mainly of old, cold petroleum. In this theory, the petroleum that was probably the most productive of chemicals to form living organisms was hot, freshly formed petroleum. The seepages that are old and established were probably very important when they first occurred.

This is a simple, straight-forward theory. It provides for the mechanism of volcanic action, the formation of the complex chemicals of petroleum, and the formation of the related chemicals intimately associated with life. Variations in chemical reactions provide a logical, consistent explanation for the diversity of living organisms.

Flora and fauna (plants and animals) of various land masses will be presented and the reader can determine which explains their presences better, evolution or multiple creations as proposed in this new theory. Chromosomes, and particularly their numbers in various species, will also be examined.

XI. THE THEORY OF EVOLUTION

It is important to distinguish the theory of evolution from theories of mechanisms that caused evolution. Darwin's theory of natural selection through survival of the fittest is an example of postulates of causes of evolution. The theory of evolution is simply that one first simple element of life having no celluar structure became a cell through gradual changes and gave rise to offspring that changed or evolved in succeeding generations to become all of the various living organisms of today. That is the basic concept and it does not depend upon theories of mechanisms that may have caused the changes. It is based primarily upon speculation that plants and animals living today are descendants of very different plants and animals that died in the remote past and left their skeletons and imprints as fossils. It is further based upon speculation that only the simplest form of life could have originated without prior living ancestors. The proponents of evolution have assumed that life originated by some process that was so complicated and so unlikely to occur that it could have occurred only once and that only the simplest form of life could have been produced.

The ways scientists have proposed life started have proven to be impossible. The proponents of evolution simply do not understand how any living organisms came into existence. They base the theory of evolution upon that lack of information, speculation that plants and animals of today descended from very different extinct organisms that left fossils, and lack of a good alternative scientific theory. The only facts upon which it is based is that extinct plants and animals that left fossils are very different than plants and animals that are living now, and the sequence of types of fossil plants and animals.

Unfortunately, the individuals of no given species have been buried in a neat sequence of layered mud. Layers of sediment are often absent in places due to erosion or lack of deposition. Fossils are really quite rare, and particularly of land animals. Even in localities where they may be considered common they are usually confined to small localities and in only certain layers of sediments. In order to present neat sequences to demonstrate evolution, fossils from all parts of the world have been mixed together in drawings and writings without any real evidence that there really was an ancestry connection.

The fossil evidence is much like finding the skeleton of a house cat that was buried in England a thousand or a million years ago and presenting it as the small ancestor of an American puma. As a similar real example, the skeletons claimed to show the evolution of horses have been found for the most part in North America. Except for recent historical times, there is no evidence that the modern horse has ever lived in either North America or South America. It has apparently never been a native of those continents. Similarly, the fossils claimed to be the primitive ancestors of camels and hippopotamuses were found in North America.

Fossils are actually mainly of plants and animals that are extinct. In historical times, many animals have become extinct, and many more are in danger of becoming extinct. When they become extinct they leave no progeny, much less an improved form. There is simply no scientific evidence that extinction of a species was different in the past than now. Once extinct, there is no evidence that any plant or animal was ever restored or replaced by evolution from any other species to fill the gap it left. The fact that fossils are of extinct plants and animals is really evidence of multiple originations of life. In order for it to be evidence of evolution, one must first assume that multiple originations of life was impossible. There is no evidence to support such an assumption.

The theory of evolution has included speculation that at times in the remote past there have been land connections between continents and between continents and islands for which there is no real evidence. That includes speculation of land bridges across the northern Pacific Ocean, between Africa and South America, between Australia and Asia, between Madagascar and Africa, and others. It further includes speculation that various select animals utilized those land bridges. The only support for those speculations is by circular reasoning. First, the theory that life could start only once and the theory of evolution must be accepted as true. Only then do the speculations of those land bridges help make the theory seem plausible. It is opinion based upon an accumulation of guesses.

Living organisms have been classified mainly based upon their structures. Classification has also been influenced by efforts to be consistent with the theory of evolution. The classification starts with division into plants and animals. From general to specific, plants and animals are then separated into phyla, classes, orders, families, genera, species, and subspecies (varieties or races).

As dramatic evidence of deficiency in geologic evidence to support the theory of evolution, of a total of twenty phyla of animals, one-half have no fossil record. There are approximately

250,000 species of plants and animals recorded in fossils covering an estimated time of 2,500,000,000 years. Very few of those species, if any, survive today by offspring even if one accepts the same skeletal structure as the only criterion for determination of species. There are approximately 1,500,000 species of plants and animals alive today. From the foregoing, one-half have no ancestors in the geologic record, and almost all of the others have only distant ancestors dependent upon accepting as distant ancestors animals with different skeletal features whose fossil remains are found on continents different than where the modern species is found.

Further, fossils found of animals simply do not verify the theory of gradual changes by evolution. Animals grossly different from predecessors appear in spurts. They seem to have appeared corresponding to changes in geologic periods or with new layers of sediments.

As for the general sequence of fossils, it can be explained as the consequence of factors affecting multiple originations. The more primitive animals require smaller lengths of DNA strand, or genome size. For example, the genome sizes of snakes, lizards, and birds are only about one-half that of mammals. As would be expected from chemical reactions and polymerizations in producing DNA, there were probably many small strands produced before there were larger ones.

Probably of more importance in determining the sequence of living organisms were pool, nutrition, and atmosphere requirements. In order to understand the cause of fossil sequence, it will be necessary to refer to specific geologic times and strata sequences. Geologic calendars start with broad general "Eras." Eras are divided into "Periods" which are determined by relatively great episodes of movements of the crust of the Earth. Lesser changes are referred to as "epochs." Important geologic Eras, Periods, and Epochs to be referred to are as follows.

		millions of years ago
	The Precambrian Eras	4,600 to 570
	The Paleozoic Era	
1.	The Cambrian Period	570 to 500
2.	Ordovician Period	500 to 430
3.	Silurian Period	430 to 395
4.	Devonian Period	395 to 345
5.	Mississippian Period	345 to 325
6.	Pennsylvanian Period	325 to 280
7.	Permian Period	280 to 225

The Mesozoic Era

8.	The Triassic Period	225 to 190
9.	Jurassic Period	190 to 136
10.	Cretaceous Period	136 to 65

The Cenozoic Era

11.	The Paleocene Epoch	65 to 54
12.	Eocene Epoch	54 to 38
13.	Oligocene Epoch	38 to 26
14.	Miocene Epoch	26 to 7
15.	Pliocene Epoch	7 to 2.5
16.	Pleistocene Epoch	2.5 to (?) possibly the last 10,000 years.

(Includes 4 glacial and 3 interglacial episodes)

When the planet Earth first formed, it is assumed that great heat was produced by compression through gravitation toward the center of its mass. The heat was the result of pressure and nuclear reactions that produced larger atoms. Because of that heat, gases were expelled which then became the Earth's atmosphere. Those gases were expelled through volcanoes before there were oceans. Because ammonia and methane are major constituents of the atmospheres of large planets without oceans, it is reasonable to assume that those gases were expelled by primordial volcanoes on Earth. Because volcanoes now are known to expel large amounts of carbon dioxide, nitrogen, and water, and because oceans in fact did form, those gases were probably early constituents of the atmosphere.

As the mass of gases cooled below the boiling point of water, the water as steam in the atmosphere precipitated. When the surface of the Earth had cooled below the boiling point of water, oceans formed. It is our basic postulate that the formation of oceans changed the basic character of the release of gases still forming from the hot interior of the Earth. The outgassing was then impeded where the crust was reinforced by the weight and pressure of overlying water. Gases then migrated under the crust of the Earth to the edges of those primitive oceans. Sediments of sand and clay had accumulated at the edges of the land masses due to the great rains eroding the land. When the gases from the hot interior of the Earth underlying the oceans reached those sediments, some of them were changed to petroleum. As more gases followed and pressures built up, the thin sediments were unable to contain them. The resulting first eruptions from those young volcanoes were mixtures of gases and petroleum, with or without magma.

As sediments became thicker and some became more compact, petroleum was produced and contained within the Earth. Where the gases could not be contained, they erupted in volcanoes. The first eruption of any volcano probably contained petroleum similar to the early volcanoes. After that first eruption, gases formed under the oceans could follow that same route to that volcano, and ensuing eruptions would have gases similar to those we now observe from established, mature volcanoes and fumaroles. Those gases are mainly nitrogen, water, and carbon dioxide. Methane is only a minor constituent, and larger hydrocarbons are not found in those volcanic gases.

Accordingly, the early atmosphere of Earth was probably composed mainly of nitrogen, methane, ammonia, carbon dioxide, and water. An abundance of chemicals of life were produced. Only certain bacteria and plants could live due to the atmosphere and temperature. Bacteria produced protein from petroleum. Oxygen was added to the air by the action of ultraviolet light on water molecules in the atmosphere. It was disseminated and mostly used in oxidation reactions, so oxygen remained too low to support air breathing organisms for a long time. Oxygen added to the air by plants was similarly disseminated, and build up of a significant concentration was slow. Plants released oxygen into the water, where its effect was more significant. The amount of oxygen that can be absorbed in water is limited.

Bacteria and plants have no "gestation" periods comparable to those of animals, so they required a pool only as necessary for the chemical reactions for their originations.

After anerobic bacteria (those requiring little or no oxygen) and plants in water were established, the stage was set for the next eruption of petroleum through seepages, mud volcanoes, or real, but juvenile, volcanoes. The petroleum again formed abundant chemicals for life, and those organisms that could live in the primordial environment thrived. That included a wide variety of aquatic animals with very low oxygen requirements. The temperature of the water was still probably a good deal warmer than it is now due to the still hot mass of the body of the Earth. Aquatic life is well known to be very sensitive to changes in temperature of water, which must be borne in mind throughout the history of the planet as it has cooled.

The precambrian, lasting from the time the Earth formed to about 570 million years ago, had bacteria, algae, and probably worms. Worms require less than one milliliter of oxygen per liter of water compared to six milliliters in modern oceans. By the time of the Cambrian the oxygen level had reached the necessary concentration in water to support shell formers. The Cambrian strata of North America alone have yielded at least 1200 species of aquatic animals including sponges, coelenterates, worms, brachiopods (bivalved shelled animals), gastropods (includes snails), echinoderms, and arthropods (crabs, shrimps, scorpions, etc.). There was a world-wide similarity in species, but already each ocean had its own species and genera. The most common trilobite of the early Cambrian, Olenellus, was extinct in the middle Cambrian. All of the 1200 species found in the Cambrian system have long been extinct.

The Cambrian period was ended with the Ordovician submergence, which caused at least half of the North American continent to be submerged under water. There was considerable volcanic activity from Alabama to New York. There are no fossils of land plants from Precambrian through the Ordovician. That is in the face of abundant aquatic life. It had nothing to do with any inability of plants to crawl out of water onto land. It is postulated that the atmosphere, possibly due to methane or ammonia, would not allow the growth of land plants. Armor-plated primitive fish appeared probably in fresh water in the Ordivician. More than 2,600 species of various shelled aquatic animals have been identified in middle Ordovician rocks in North America. True corals first appeared. The most distinctive and common animals of the Cambrian were trilobites. The most abundant and distinctive of the Ordivician were graptolites. Clams also became abundant.

During the Silurian Period, most of the land again emerged from the ocean. Volcanoes were active in the northern part of the Appalachians, Alaska, and Northern California Mountains formed in England and Scandinavia. Marine invertebrates still dominated. Sea scorpions, eurypterids, were common in only this Silurian Period. One species had a body seven feet in length, the largest arthropod of all times. Very primitive fish left rare fossils. A few land plants appeared. Rare millipeds and scorpions less than three inches long may have been air-breathing animals, the first evidence of such.

With new organisms originating and teaming in the oceans for more than 140 million years, and a sea scorpion with a body seven feet long, the appearance of a few primitive plants, a miliped, and a three inch scorpion on the land can hardly be considered evidence of evolution. Finally, the atmosphere had apparently changed enough to permit at least a few primitive plants and possibly primitive animals that required little oxygen to live on land. There is no evidence of any deficiency in ability of new organisms to originate at any time. To the contrary, the only apparent limitation on varieties of living organisms was the harsh environment.

During the following Devonian Period, uplift of the Appalachians was renewed dramatically. It formed the second generation of those mountains. In England, the formation is known as the "Old Red Sandstone." That has particularly significance concerning oxygen. The red color is due to oxides of iron which required oxygen to produce, either locally or generally. That indicates considerable oxygen had been produced and again combined to be unusuable for animals. That red sandstone is the first good evidence of oxygen of significant concentration in the atmosphere.

Corals and brachiopods were abundant in the Devonian Period. Fish became abundant and diversified in seas, lakes, and streams. There were sharks twenty feet long. Air-breathing lung fish appeared. The African lungfish, presently existing, can go into mud and breathe air during dry seasons and then return to breathe in water during wet seasons. An Australian species can come to the surface and breathe air when the water is stagnant.

The first land vertebrate known appeared at this time. It was about three to four feet long, shaped like a crocodile, and was found in Greenland. It is believed to have been an amphibian.

Land plants were abundant for the first time. Trees were primitive scale trees. Primitive seed ferns were abundant. They constituted the first forests. In Scotland, associated with plants were 3 genera and 18 species of spiders, 1 form of mite, and a wingless insect.

None of the foregoing is evidence of evolution. It clearly reveals that whatever the source of living organisms was, that it was an abundant source. It was a source that provided new species as soon as the environment would support them. As soon as plants could survive in the atmosphere, animals that had low oxygen requirements were also produced. But, as will be shown with a presentation of Australian phenomena, it was necessary to have crustal movements and volcanoes to have such a source of abundant life. In this theory, the plants and animals were produced as a result of crustal movements that produced mountains and new juvenile volcanoes emitting petroleum. Whatever was poisoning the atmosphere for plants in the Precambrian, Cambrian, and

Ordivician had apparently been adequately removed, and with primitive ferns and trees growing in abundance, oxygen had been produced in sufficient quantities to allow some of the new originations of life to live on land and new originations that lived in water. It is noteworthy that in this one geologic period which reveals the first evidence of significant oxygen in rocks, vertebrate fish, a vertebrate amphibian, land plants, and land insects were added to the flora and fauna of Earth. That was a time period of 50 million years after 175 million years of invertebrates confined to the sea. The atmosphere was in a delicate balance, with volcanoes adding gases and plants now producing oxygen.

In North America, the following Mississippi Period was marked by more widespread disturbance than any previous period. Mountains were made in Appalachia, Colorado, Arkansas, and Oklahoma. Europe produced the Variscan Mountains in Ireland, France, Switzerland, Germany, and Austria. There was much associated volcanic activity.

Trilobites were almost extinct. No insects have been found in fossils of the period. Fishes were less varied than in the Devonian. It was a time for some apparent extinctions. Land plants remained abundant. Amphibia have been found in Europe, and tracks of similar animals have been found in West Virginia.

The Pennsylvania Period was marked by submergence of 30 percent of North America under water. Uplift continued in Oklahoma, Colorado, and Texas. At the end of the period the Continent had again emerged out of the sea. More mountains formed in Europe and in the Himalaya region. Fossil ferns and trees were large with abundant foliage. None were like modern plants. Trees were spore bearing and ferns had small nut-like seeds instead of spores as modern ferns do.

Insects were larger than at any other period in the Earth's history. Many were more than four inches in length and some exceeded a foot in length. A dragonfly-type of insect had a wingspan of 29 inches. There were centipedes, spiders, scorpions, and about a hundred types of amphibia. A reptile fossil of the period has been found. Land snails first appeared.

This Pennsylvanian Period was noted for the huge amount of coal formed. The reason the trees, including even their leaves, and the ferns did not rot to carbon dioxide and water was probably due to the continuing low concentration of oxygen in the atmosphere. The burial of those huge forests without their consuming oxygen by rotting added to the total effect of the oxygen they liberated into the atmosphere.

In the following Permian Period great Paleozoic geosynclines (troughs that had been filled with sediment) were uplifted to form great mountains located along the eastern and southern borders of North America, across central Europe and southern Asia, and eastern India. The Urals were also formed. The Appalachians were probably as high as the modern Alps. Volcanoes were active in California, western Nevada, western Idaho, and eastern Oregon. Volcanic deposits in the latter areas were thousands of feet thick. There was also much volcanic activity in central Mexico and Alaska. Deserts became more widespread than at any time other than the present. Glaciers formed over wide areas even approaching the equator in the Southern Hemisphere, including in South America, South Africa, Australia, and India.

By the close of the period, the trees and ferns that previously grew and produced the Pennsylvanian coal were nearly extinct. True conifers appeared and rapidly dominated as woody trees. Insects were small but varied and the large ones were extinct. Mayflies, true dragonflies, and beetles appeared. Amphibians thrived and reptiles up to ten feet in length appeared. Most brachiopods, all trilobites, some corals, most crinoids, all blastoids, and two orders of bryzoa became extinct. With the species added, total Permian ocean life was still abundant.

The Triassic Period is characterized by three layers of sediment that are prominent in the Colorado Plateau, including Zion National Park and the Grand Canyon of the Colorado. The lowest layer is a deep red, the middle light tan, and the top layer is a greyish, light green. The lowest layer is also red in Great Britain and Europe indicating a general similar oxygen concentration. Petrified wood common in the upper layer suggests that oxygen was still inadequate to cause the wood to be readily oxidized to carbon dioxide and water.

Lava flows were immense in the Pacific Northwest and South Africa. Ash was scattered from western volcanoes in North America. Some movement occurred in the Appalachians which were minor compared to previous periods.

Seed ferns and trees of previous periods had nearly vanished. There were swamps with ferns in the southern part of the United States, and large conifers in the West. Reptiles were common and varied. Dinosaurs arrived. Dolphin-like reptiles appeared in seas. The most common marine shelled invertebrates of previous periods, ammonites, became nearly extinct.

It is to be noted that the oxygen supply in the atmosphere was still probably only adequate for cold-blooded reptiles and insects.

The new conifers had probably not become established forests that compared with the oxygen production of the huge, dense forests that produced the coal of the previous Pennsylvanian period. Oxygen was continually being consumed by iron on the surface of the Earth.

In the following Jurassic Period, redbeds attained a thickness of 3000 feet or more in the Colorado Plateau. They formed the massive Wingate and Navajo cliffs that are prominent in Zion National Park and the Grand Canyon. They are topped by the Morrison formation which is noted for dinosaur fossils.

At or near the end of the Jurassic Period, the Nevadian disturbance occurred. The Pacific Coast geosynclines were crushed by forces from the west and the thick deposits were thrust upward to form long ranges of mountains in British Columbia, Washington, Oregon, California, and Utah.

The Wingate and Navajo are essentially without fossils. The Morrison formation has yielded 69 species of dinosaurs, 3 crocodiles, 24 river claims and land snails, 23 plants, and 25 primitive mammals. The Jurassic Period was an important coal producing system in Siberia and Australia, again indicating deficiency in oxygen.

Up to this point it appears that the limitation of species has been the atmosphere. DNA productions had produced a variety of bacteria, algae, seed-bearing ferns, spore bearing trees, invertebrate shelled animals, and more recent batches of DNA provided vertebrate fish, amphibia, reptiles, and conifer trees. A batch in the Jurassic apparently produced tiny marsupial mammals. The crop was probably limited due to still deficient oxygen supply. More than 200 million years had elapsed since the first amphibia had been produced.

For originations or creations of plants, shell creatures, and reptiles, pools were probably required for the chemical reactions to produce germ cells and to provide nutrients and oxygen for the relatively short incubations or gestation periods. At the ends of the gestation periods, the reptiles would crawl out of the water the same as they crawl out of shells.

Marsupials would require essentially the same type of pool with perhaps more oxygen, and they would require more warmth and stability. Apparently such pools were available in the Jurassic. Other mammals would require pool stability for much longer periods of time. Survival of newborn mammals and many birds would have probably required adoption by other established species.

Petroleum from juvenile volcanic eruptions onto rocks and into pools of water, particularly associated with fumaroles and warm areas, probably formed pools adequate to produce plants, shelled animals, insects, reptiles and birds as evidenced by creations or originations of them on volcanic islands such as the Hawaiian Islands and the Galapagos Islands which will be presented. Placental mammals, and particularly humans, may have required more stability and more nutrients available in petroleum seepages.

In addition to the arrival of marsupials, modern types of insects appeared in the Jurassic including flies, dragonflies, beetles, grasshoppers, moths, termites, and ants. Cockroaches had arrived earlier.

The first birds are claimed to have appeared in the Jurassic period. The authenticity of the fossils is questioned. In this new theory of multiple originations or creations, presence or absence of so-called "missing links" makes little difference. Something appearing to be a missing link would probably be just another unusual combination of parts possible in DNA production. This presentation of fossils and geology is made to show that the sequence of fossils can be explained without evolution. The arrival of birds makes a difference only in establishing the time oxygen reached a concentration for their needs. Their oxygen needs are greater than those of humans which is the reason miners have traditionally taken birds with them into mines whose deaths warn them of gases that displace oxygen. The problems caused by counterfeit fossils increases in importance in later periods, so the matter should be addressed at this first significant probable fraud.

The first bird is generally claimed to be the Archaeopteryx. A feather imprint was claimed to have been found in a quarry of fine grained limestone used in lithography in Solenhofen, Bavaria, about 1860, by a politician physician, Karl Häberlein, the district health officer. The Pterodactyl, which is the reptile equivalent of a bat, was found in the same formation in 1784. Thomas Huxley, the great advocate of evolution, had drawn a sketch of hypothetical primordial bird which was feathered, had claws on its wings, the teeth of a reptile, and a lizard-like tail. It may be noted that by usual classification, the teeth would most likely establish such a creature as a feathered reptile. Almost immediately after Huxley's prescription, such a fossil fitting the description was found in the same quarry as the feather imprint was reportedly found. Häberlein purchased the new find. He then allowed scientists to examine it under strictly controlled conditions, including no notes being allowed. The creature was pigeon-sized, about the size of a pterodactyl, the head was missing, and an extremely long lizard-like tail was feathered on both right and left sides with large, vane-shaped feathers. Most naturalists proclaimed it to be a fraud immediately. After a drawing was published from memory by a museum curator, Häberlein found a buyer. The British Museum was persuaded to buy it. The museum paid 600 marks and declared it to be the first bird. There was still the matter of predicted reptilian teeth. Sixteen years after that find, an owner of a quarry near the original find, found one with a head, including reptilian teeth. The quarry owner sold it to Ernst Häberlein, son of the deceased physician. It was then offered for sale for 36,000 marks. No one could raise the money. Appeals were made by evolutionists to various governments to purchase it. Finally, Werner von Siemens, a German industrialist, purchased it for 20,000 marks to retain German pride in the find, and then sold it to the Mineralogical Society of Berlin University. In 1956, another one was found about 200 meters from the first find, again without a head.

Fossils have had a good market and have had prestige value to those who have found them for centuries. In Würzburg, in 1724 to 1726, a number of fossils were counterfeited in a conspiracy contrived by two professors to discredit and embarrass another professor at the same university. It included three adolescent boys who helped plant and excavate the "fossils."

Near the bird finds, on Lake Constance, in the town of Öhningen, quarries owned by the Church were rich in fossils. There was a good market for the good and unusual ones. From 1775 to 1850, the project became a forgery workshop. It started with piecing fragments of bones and leaves together. New genera and new species were fabricated. Bones of horses and pigs were imbedded in moist limestone. Cements, weathering agents, and stains were cleverly used. An alligator-tortoise composed of fragments from six animals sold to the Teyler Museum in Haarlem resulted in exposure of the frauds.

Piltdown man was fabricated by prominent scientists in a similar fraud. It involved an ape's jaw that had been carefully altered. Skill in taxidermy has apparently been as important as knowledge of taxonomy. There have undoubtedly been many such frauds in the zeal to establish the theory of evolution over other concepts of life.

The bird fossils are particularly suspect for several reasons. It was predicted by the most prominent evolutionist as a "missing link" shortly before the find. A century earlier pterodactyls, about the same size, were found in the same formation. It would be unusual to have two such creatures in the same neighborhood at the same time. No other reptiles have shown any tendency to have feathers and no existing birds have teeth. There is no logical reason in the theory of evolution for birds to have lost teeth in all species. Replacing a jaw with a bill would not be any easy step in that theory. No bird in existence has a tail of the same type or tail feathers in the same pattern. The entire marketing story by a physician-politician (politicians of any sort are not noted for integrity) is suspect compounded by its occurring at a time and place where counterfeiting of fossils was an established lucrative practice. Furthermore, they are too rare to have been the ancestors of modern birds. They would surely have had to reproduce a great number to complete the postulated remaining evolutionary transitions.

A departure was made from usual evidence submitted in support of the theory of evolution. Usually species are claimed to have some unknown, undetermined, "common ancestor" that cannot be examined. The stem of family trees is missing, only the branches are identified. Unanticipated, chromosome studies, to be presented in detail, have proved that common ancestors of different species are impossible.

The Cretaceous Period marked the end of the Mesozoic Era. It was dramatic. The Laramide Revolution was the most widespread movement of the crust of the Earth in its history. The Rocky Mountains and the Andes in South America were produced. The Alps and Himalayas were increased. Volcanic activity was present throughout the western part of the United States. Huge coal deposits were produced. The forests became essentially modern with deciduous trees and conifers. Grasses, cereals, seed and fruit-bearing shrubs, flowering plants, and our common vegetables arrived. Two species of birds with teeth were found in Kansas. One had vestigial wings that made flight impossible. A head of a bird without teeth was found in Alberta, Canada, from the same period. Mammals included marsupials and insectivores. All dinosaurs became extinct as did all winged reptiles. The great marine reptiles all became extinct except for the turtles. Snakes first appeared. Several types of shelled invertebrates also died out.

None of the extinctions can reasonably be explained by the theory of evolution. None were due to any apparent problem with competition or lack of evolution. A few small primitive mammals that had arrived in the previous period were certainly not competitors with the dinosaurs. In contrast with the destruction and extinctions of plants which formed coal, extinction of the large terrestrial and aquatic reptiles, and extinction of shelled aquatic invertebrates, all sorts of modern types of plants as well as birds originated.

In the presented new theory, widespread crustal movements and new volcanoes provided new batches of DNA. New arrivals were new originations of life. There is no reason to believe that any were the progeny of very different plants and widespread animals that suddenly became extinct on all continents.

The mammals and birds that originated in the Cretaceous also became extinct, and coal was deposited from time to time, so oxygen concentration probably remained a critically, delicately balanced factor for some time. Recalling that the primordial atmosphere was probably poisonous to land plants, as there were no such plants until long after complex creatures were swimming in oceans, volcanoes from time to time probably have emitted similar substances affecting the lives of terrestrial plants and animals.

The Cenozoic Era has been one of continuing growth of mountains, especially the Cascades in North America, and the Alps and Himalayas in Europe and Asia. Volcanic activity has been pronounced worldwide.

Mammals of the Paleocene Epoch included insectivores as holdovers from the Cretaceous and Jurassic. Dominant mammalian groups in the Paleocene were the now extinct large members of three "archaic" orders. They were condylarths, amblypods, and creodonts. There were also small rodents and primates.

The Eocene was characterized by small animals believed by evolutionists to be precursors of presently existing animals. They included dog-sized horses, small hornless rhinoceroses, tiny camel-like animals, squirrel-like rodents, bats, and primates. The largest were two to three feet tall. Large archaic mammals also existed which were dog-like, cat-like, and hyena-like. Others were much larger. Sea mammals, now also extinct, fifty feet long appeared.

By the oligocene epoch, most of the archaic types had become extinct. There were now beaver, squirrels, rabbits, mice, dogs, and cats. Early primates had become extinct in North America. There was only one ape species in Europe.

The Miocene featured pony-sized horses with side toes, sheep-sized camels, abundant oreodons, rhinoceroses, giant pigs, rodents, dogs and cats.

Pliocene fossils are poorly preserved. Mastodons appeared.

In the Pleistocene, there were four large species of elephants in North America. The Pleistocene, the past two and a half million years, has experienced four glacial and three interglacial periods. All of those types of elephants are extinct. If one is willing to accept the postulate that modern horses were dog-sized with different kinds of feet, then those animals did not become extinct. If one is willing to accept the postulate that animals the size of sheep became camels, then they did not become extinct. There is really no more reason to accept such a postulate than there would be to say that a housecat is the same animal species as a lion. Probably most modern species of mammals have not been in existence much longer than humans, or they may have originated at essentially time.

Of twelve families of birds in the Cretaceous, only two exist. Most modern birds probably originated in the Eocene. Large birds appeared on all continents, and most became extinct.

Sequence of fossils up to the Cenozoic was probably determined by atmospheric changes. At first it was probably poisonous. Later the determinative factor was apparently the oxygen concentration. Those atmospheric changes probably account for most of the gross changes in types of fossils in various geologic systems claimed by others to have been the result of evolution.

Recurring extinctions and arrivals of new mammals during the Cenozoic is probably the result of the most dramatic changes in this Era. It is a period of recurrent glaciation with changes in temperature and crustal and volcanic activity. Each batch of DNA was produced under slightly different conditions. There has probably been a gradual change in oxygen concentration and temperature of the Earth during this period in addition to the recurring episodes of harsh changes associated with glaciation. Earth was hot all over and has been continually cooling.

Certainly the fossil sequence can be explained on a scientific level based upon changes in chemicals for life's originations, temperatures, atmosphere content including volcanic gases and oxygen, stability of the crust of the Earth, and available nutrition. To explain differences in modern species and those of fossils there is no need to resort to a theory of evolution that has no basis in scientific fact or reasoning.

An interesting example of sequence of flora and fauna similar to geologic history that was not determined by evolution was provided by the volcanic explosive eruption of Krakatoa in 1883 which wiped out all life on the island. Three years later its flora consisted of blue algae, 11 species of ferns, and 15 species of flowering plants. Six years after the eruption, fauna had been added consisting of spiders, butterflies, and monitor lizards. In 1906, it was found to have 14 species of plants, 40 arthropods (insects, etc.), 4 snakes and 2 other reptiles, and 16 species of birds.

Fossil records do not reveal one fountain where evolution occurred and from where plants and animals spread. Each continent has fossils. When two continents have fossils even grossly similar, it seems that every effort has been made to connect the two to a common source from which there could have been spread. This is to make nice sequences to fit the theory of evolution. The jumps observed on each continent are otherwise inconsistent with the gradual changes the theory requires. When such spread is completely unbelievable, separate but parallel or convergent evolution has been given as an explanation. That assumes that after hundreds of millions of years of separation, that some primitive animals changed to produce essentially the same animal on different continents. That is inconsistent with inability to replace extinctions.

Usual logic is reversed. The most logical starting point is that different species started out differently. The only relevant fact we can observe and verify scientifically is that a given species can only be a distant descendant of the same species.

If that does not explain all relevant observations, the logical sequence would be to postulate that evolution took place everywhere with present similarities of flora and fauna being the result of similar chemicals to start with and exposure to the same type of environment. It may be noted that the most ancient of fossils of the lowest layer of sediment, the Precambrian, have been found in Canada, South America, Africa, Australia, Greenland, the British Isles, the Scandinavian Peninsula, the Arabian peninsula, and other parts of Asia. Living organisms apparently became abundant all over the Earth, in seas around all continents, at about the same time. The theory of evolution requires extreme pliability of all plants and animals to account for essentially all species to have resulted from very different fossil ancestors. With evolution as postulated, early living organisms could have developed into a complete flora and fauna on all continents. Unfortunately, the environments including flora and fauna on the various continents in general as well as in specifics are not similar enough to account for the similarities required in some species after the hundreds of millions of years of separation and still account for dissimilarities in other species and absences of entire families and orders.

The proof required to accept such evolution as having occurred should be greater than that required to accept separate origins. As noted, the only real observation is that species only reproduce their own kind over repeated generations. The concept of evolution has not been verified by direct observation and is only speculative.

Subject to greater proof requirement is the conclusion that two similar animals on different continents came from the same source. Migration cannot just be assumed. There are not just water barriers, which are tremendous, but there are also temperature, mountain, tundra, snow, ice, desert, predator, and competitor barriers.

As it is even more improbable, greater proof should be required for the conclusion that two dissimilar animals on different continents, sharing a few gross physical traits in common, came from an ancestor with only crude similarities in selected features. The latter type of conclusion appears to be the rule and not the exception. It requires the most imagination and speculation, and has no supporting evidence. It is based strictly upon assumption of evolution as fact and that the existence of all plants and animals must be in accord with that theory. It is an example of circular reasoning. No scientist would predict beforehand the evolutions claimed to have occurred. Any scientist who would predict, for example, that a shrew would change through generations to become a human being would be considered to be completely incompetent, to say the least. Surely, no scientist disciplined to depend upon observed, repeatable facts would predict that descendants of a fish would grow feathers, sprout wings, and fly. The lines between physics and metaphysics and between science and science fiction are blurred.

Other observations claimed to support the concept of evolution relate to embryology. The developing human embryo has certain gross characteristics and organs in the first few weeks of life that resemble those of a fish. None of the fish-like organs are functional or ever become become functional. If embryology reflected evolution, the more recent would be expected to be more conspicuous than the postulated very remote. There is no claim that at some period the human embryo resembles a monkey, a lemur, shrew, salamander, or other animal claimed to have been an ancestor. If a fish period is claimed, it would be only reasonable that some of those other periods would be required. To be consistent, an interruption of gestation by abortion, miscarriage, or early delivery should occasionally result in a living fish or some other creature claimed to be represented in embryologic development by recapitulation of evolution. No embryo of any mammal has ever developed into a fish, salamander, or any type of animal other than the species of its parents.

The pattern of differentiation, development, growth, and maturation of an embryo and fetus are probably part of the genetic code and there is no proof that it has any relationship to prior history of changes in ancestors. It could relate to certain required combinations of nucleotide sequences in DNA. There is no evidence that any history of an animal is transmitted to the DNA of its sex cells, the ova and sperm, to be transmitted to offspring. That is required for embryology to have any relevance to the theory of evolution.

As noted above, the only real relevant observed and observable, repeatable fact is that one species cannot produce offspring of a different species that is fertile. That fact has stood the test of more than one hundred years of intensive efforts by scientists of all disciplines to prove it to be false. Thousands and perhaps millions of generations of fruit flies have been observed in laboratory situations. No gnats, mosquitos, bees, ants, or other animals have evolved from those flies. Other laboratory animals, farm animals, and humans have also shown no changes of evolution.

Fossils are more supportive of the concept of multiple originations of living organisms than they are of the theory of evolution. They are evidence of many extinctions and arrivals of organisms of varying degrees of differences from predecessors. Succeeding organisms have also become extinct. For the most part, different types of animals are living now than those represented in even recent fossils. There is no evidence that species can give rise to different species or different genera. With those factual premises, the logical conclusion is that the plants and animals living now that are different than those of fossils either lived in the remote past and left no fossil evidence or they originated after the extinction of those that left fossils. If life originated in the manner postulated in this new theory, it is probable that most organisms living now originated after the extinctions of those different animals that left fossils.

XII. THEORIES OF CAUSE OF EVOLUTION

The reason, force, or cause that resulted in postulated evolution is generally accepted as having been natural selection by survival of the fittest and non-survival of others. On the surface the theory of natural selection by survival of the fittest appears to be reasonable and logical. As an example, if an animal could improve by running faster, jumping higher, building bigger muscles and the like, it and its descendants that continued to so improve would be more likely to survive and its competitors that were not as athletic would be less likely to survive. Repeated improvements would add up to a different species, and even a different genus, family, etc.

Two basic problems are readily apparent. One is the problem of getting the message to the gonads. The other is whether or not natural selection by survival of the fittest has actually functioned to cause evolution.

Darwin recognized the problem of transmitting acquired characteristics to succeeding generations. It requires getting the message to the cells which produce offspring such as ova and sperm. To transmit the information to descendants, Darwin avoided changes in sex cells and proposed that "every unit or cell of the body throws off gemmules or undeveloped atoms, which are transmitted to the offspring of both sexes, and are multiplied by self-division . . . their development into units or cells, like those from which they were derived, depends on their affinity for, and union with other units or cells previously developed in the due order of growth." *The aggregate of his "gemmules" is akin to a miniature replica of the body circulating in the blood. There was and is no supporting scientific evidence for the hypothesis. It is considered fanciful and has been discarded by most scientists.

The use-disuse hypothesis was the first proposed cause of evolution. It was theorized that if a body part was not used that offspring were born with that part defective. After repeated generations of such disuse, offspring were born without the part at all. Tails were cut from rats of succeeding generations whose offspring continued to be born with tails. Convinced by the lack of logic after the tail experiments, the disuse part of the theory was discarded by most scientists. It is apparently still accepted by those who appear on television and write textbooks. The use part of the theory is even more tenacious. If a fish used its fins to move about on the banks of streams, according to the theory, nature accommodated by providing offspring born with legs. How

*Darwin, p. 584

the message reached the ova and sperm is simply ignored. Obviously, the concept is not based upon scientific observation.

With acceptance of the use-disuse idea tenuous, another explanation was forthcoming. It was observed that offspring of plants and animals occasionally experience a mutation. It can be an abrupt change. An animal may be born with some part that is quite different than the corresponding part of its parents, or a plant may be different in some respect from its ancestors. It was postulated that if such a chance mutation occurred that made an animal or plant more fit to survive, its progency would survive and its unchanged relatives would become extinct. If the process was repeated over succeeding generations for hundreds of millions of years, it has seemed to be a reasonable postulate that new species would result, new genera would result, and so forth. That is the generally accepted theory of the cause of evolution.

A problem was soon recognized. In order to preserve and benefit from a mutation the individual and subsequent individuals with that particular mutation must be isolated from relatives. Otherwise, the mutation would be diluted back into the same reservoir of genes and no significant effect on survivability would result. Ostracism could have that effect, but such behavior related to variants is apparently not practiced extensively except by Homo sapiens, the human species. Sexual attractiveness is another possibility for separation of mutants. That would require the mutation to make the individual more appealing to the opposite sex with more offspring as a consequence. In nature there is really no evidence that those less attractive have difficulty in procreating. Isolation of mutants can easily be postulated without evidence, so the problem has not had significant influence.

A more serious problem with the theory of evolution by mutations has been learned. Over 99% of the mutations observed are harmful. Of millions of generations of flies, mice, rats, cats, dogs, monkeys and other laboratory and farm animals studied, apparently no structural mutations have been observed that are advantageous or point toward evolution of animals with characteristics making them more fit to survive than their unchanged siblings. Mutations go in the opposite direction. They are destructive. They make animals less fit to compete. Fruit flies can be born without wings, animals can be born without limbs, but none have been observed with a new part that is beneficial. Many humans have been born without vision, without hearing, or without limbs. They have been born mentally deficient and sterile. No one has seen a human born with a part of his body that is structurally different which is superior to that of the species. No snakes or fish have grown legs.

With the imminent demise of the theory that mutations caused evolution, other mechanisms have been proposed. Genetic drift has been postulated. That is the concept that an isolated small population may lack representatives of the total complement of genes of the species. With time it results in observable differences from ancestors. That is back to Darwin's original work. He showed that all fancy varieties of pigeons could have been derived from the common wild type we see in parks. It could have been accomplished by separation of individuals on the basis of some prominent individual characteristics. When the fancy pigeons were mixed together, they reverted again to the wild type. There were no origins of new species shown. There were only subspecies, varieties, or races. Origin of new species was speculated by Darwin without scientific evidence.

Another postulate is that evolution is concerned with only form of the body and not genes. It proposes that plants and animals change to fit a niche in the environment without any change in genetic make-up. According to that theory, there are niches for all known species that can be filled by any type of animal that is available. Body type is postulated to change. Reduced to an example, if niches cannot be filled for cats and dogs by members of those families, opossums will change to look like cats and dogs to fill those niches. The postulate has no scientific proof and denies the known facts and principles of genetics.

A related postulate conforms to the interpretation that evolution has occurred in spurts probably related to major disturbances in the environment. It proposes that under great stress upon parents that developing embryos change to counter that stress. Repeated over generations, the form of the animal is postulated to have changed to a different species. That theory also does not conform to observations in genetics which correlate body structures with specific chromosomes.

It is submitted that theories of cause of evolution are quite imaginative. There still remains the ultimate question of whether or not they were actual causes of evolution.

All of the theories of causes of evolution are variations of natural selection by survival of the fittest. Theories of changes in body structure because of use or disuse, changes in genes by mutations, changes in body form to fit niches, and changes in embryos are all directed to changing an organism to make it more able to compete or survive in the environment. They depend upon filling non-competitive niches and/or having competitors die.

From the broadest overview, it is difficult to imagine any type of natural catastrophe in which the most primitive of plants and animals would not be more likely to survive than the more complex large organisms. Fish, insects, bacteria, and algae are far more likely to survive than organisms that have appeared later.

There is no reason to think that the extinct mammoths and mastodons were any less evolved than elephants which survived. There is nothing about sabre toothed tiger skeletons which supplies a reason for their demise in a region where the cougar, or American puma, now lives. From their fossils, there is no reason to attribute extinction of any plants or animals to an evolutionary status.

There are inherent difficulties in evaluating the concept that survival was due to some advanced change in a plant or animal. The whiteness of the polar bear is claimed to have given it a survival advantage by camouflage. Perhaps a black bear would have been better able to keep warm. After-the-fact type of analysis has opposite rules that serve conveniently when predictions are not required. If an animal is dull in color, it is claimed it survived because of camouflage. On the other hand, if the animal is brightly colored, its survival is attributed to increased sexual attractiveness or that the color served as a warning to enemies. It is impossible to predict extinction of any plant or animal based upon its low position by evolution standards.

We can only speculate upon what actually has caused extinction of plants and animals in the past. At the end of the Cretaceous period 65 million years ago, it has been claimed that at least 75% of the species including dinosaurs were wiped out. Approximately 50% of species of flowering plants and marine organisms became extinct. In a disaster of such proportions, the proposal has been made that it was not a matter of fitness but more aptly described as "survival of the luckiest."*

Diseases could probably cause extinction of species of plants and animals. Insects carried from one region to another are a constant threat to agriculture. An animal arriving on a different continent might carry germs to which native plants and animals have no immunity. Survival would not depend upon any prior changes in the postulated process of evolution. It would depend mainly upon chance exposure to the germs.

No evolutionary process has made succeeding generations more able to survive a volcanic explosion accompanied by large amounts of gases and ashes. The development of quick mobility

*Hitching, pp. 166-167

might help avoid small flows of lava, but there is no reason to believe that that type of event was anticipated in the postulated evolutionary process.

Adaptation to drought or famine might be related to survival. Mobility and movement or spread to a more favorable environment by those not prior suited by chance to such environment is more reasonably related to survival in such a situation than change in structure. Droughts and famines in human experiences are effective in eliminating the non-mobile before effective evolution would have time to occur.

The many layers of sediments covering the continents is evidence of repeated flooding. It is inconceivable that the postulated evolutionary process has been geared to surviving such recurrent catastrophes.

The mountains of Earth suggest that there have been earthquakes whose severities have been beyond our imaginations. There is nothing in the postulated evolutionary process that has made succeeding generations better equipped to live through such shocks.

There is no real evidence that survival of any plant or animal has been due to its more advanced status from an evolutionary viewpoint. By the same criteria, there is no evidence that any of the fossil plants and animals became extinct because of their being more primitive than those that are living today. Complex and specialized organisms are actually probably more susceptible to extinction than the more primitive organisms.

In summary, the theory of evolution and the postulated causes of evolution are completely speculative. They are inconsistent logically and their premises are in conflict with the facts. There is no evidence that favors the theory of evolution over the theory of multiple originations. The simplest, most logical explanation for the variety of living organisms is that of multiple originations or creations of essentially every species and some subspecies. The presented new theory provides for an abundance of source materials and frequent ideal situations for living organisms to come into existence. Observable scientific evidence in support of the new theory and in conflict with the theory of evolution follows.

XIII. CHROMOSOMES AND PROOF OF NON-PATERNITY

Except for the simplest plants, the cells of plants and animals are divided into nucleus and cytoplasm. The nucleus is surrounded by a thin membrane which separates it from the rest of the cell. Cytoplasm surrounds that membrane and its enclosed nucleus and the cytoplasm in turn is surrounded by another membrane which marks the boundary of the cell separating it from other cells and body fluids which occupy spaces between cells. The DNA which contains the genetic code is in the nucleus. It is divided into a number of parts of different sizes, shapes, and lengths. Those parts are called chromosomes. The number of those chromosomes, their sizes, shapes, and lengths are specific for each and every species of plant and animal. When cells divide the chromosomes divide so that each of the two resulting daughter cells receive a full complement of chromosomes. When cells which form germ cells divide, the pairs of chromosomes separate so that each sperm or ova receives one chromosome of each pair, with a total of one-half of the number of chromosomes of a cell of a parent. Then when the sperm and ova unite, the resulting cells have the same number of chromosomes as the parents. The chromosomes of the sperm must pair-up with the corresponding chromosomes of the ova. They must match. That is the reason that only animals of the same species can produce fertile offspring. The same system operates in both plants and animals and prevents species from giving rise to different species.

The number of chromosomes found in a germ cell is called the haploid number and the number found in other cells, usually twice that number, is the diploid number. Some plants and some fish occasionally triple and quadruple the haploid number of chromosomes in their cells. It may occur by male or female germ cells that had the full parental complement of chromosomes instead of the haploid number. Multiple numbers of the haploid number of chromosomes does not change the species. If it occurs in mammals, the embryo does not survive.

In the human there are individuals that have chromosome numbers different than the usual 46. Chromosomes are numbered based upon the characteristic size and shape of each There is an extra chromosome number 21 in Down's syndrome, making a total of 47 chromosomes instead of 46. The resulting person is characterized by mental retardation, slanting eyes, short stature, short, broad hands with crease patterns similar to those of monkeys and apes, and other deformities. An extra number 13 chromosome, total of 47, is found in Patau's syndrome. The person having such a condition is characterized by mental deficiency, deafness, cleft lip and cleft palate, polydactyly (too many fingers and toes), heart deformities, and prominence on the back of the heel. An added number 18 chromosome, total 47, is present in Edward's syndrome. A person having that syndrome has multiple congenital defects of various organs, low set malformed ears, receding mandible, and mental deficiency. Ninety per cent die within the first six months after birth.

The usual human female has two female sex chromosomes, labeled "X" chromosomes. The usual human male has one female or X chromosome and one male chromosome which is named the "Y" chromosome. A female that has only one X chromosome, with a total of 45 chromosomes, has Turner's syndrome. Such a person is retarded in sexual development and is usually sterile, is short in stature, has webbing of the skin of the neck to the shoulders, heart deformities, and hearing impairment. Males may rarely have in addition to their usual X and Y chromosomes one to four more X chromosomes (total 47-50) resulting in Klinefelter's syndrome. Such a male is subfertile with small testes, is mentally deficient, has developed breasts, and other abnormalities. A female with an extra X chromosome, total 47, is frequently mentally retarded and has underdeveloped genitalia and subfertility. An extra Y chromosome in males, total 47, is associated with excessive stature, mental deficiency and antisocial behavior.

Sometimes two chromosomes may fuse, resulting in a decrease in numbers without change of species. When considering possible changes in chromosome numbers over hundreds of millions and billions of years, that is taken into account by counting limbs or segments of chromosomes.

Findings of chromosome numbers, sizes, and shapes devastate the theory of evolution. Even counting the segments and limbs of the chromosomes does not save it.

Genetics may not always determine parentage, but it often proves that one animal could not have been the parent of another. The theory of evolution requires actual descent as from a parent to offspring repeated millions of times. If all descended from one parent, all plants and animals should have cells with the same number of chromosomes having the same size and shape. Not even species that have just minor apparent differences have the same number, shape, and size of chromosomes. Chromosome numbers, sizes, and shapes simply do not correlate with the theory of evolution.

As examples, the horse family includes the donkey with 62 chromosomes, the Mongolian wild horse with 66, the domestic horse with 64, and three species of zebra with 32, 44, and 46 chromosomes. A common ancestor is impossible. Bats of 12 different species have 10 different diploid chromosome numbers between 10 and 50. Four different species of armadillos have diploid chromosome numbers of 58, 60, 62, and 64. A cottontail rabbit has 42 chromosomes, the laboratory rabbit 44, snowshoe hare 48, and a jack rabbit species has 48.

It is noted in comparing chromosome numbers and the number of their parts that although parentage and lineal descent from common ancestors is disproven, animals that are similar in appearance and are classified by structure as being of the same genus not uncommonly have chromosomes that have general similarities in size, shape, and structure. That is not unexpected in the new multiple originations theory. This theory bases similar animal structure upon similar DNA. It postulates that the DNA came from common chemical pools with large parts of the DNA strands repeated, but with variations. Some variations include different combinations of parts of DNA strands. The same general types of chemical combinations have occurred at various times and places throughout the world. Regardless of time and place it was formed, DNA of similar animals is expected to usually be similar in size and similarly divided.

Primate chromosome studies are of special interest. As noted above, alteration in numbers of human chromosomes has caused serious deformities. No evidence toward an improved species has been reported with either the addition or subtraction of chromosomes.

The human, *Homo sapiens*, (family Homidae), has 46 chromosomes, 44 being autosomes (body types) and 2 being sex chromosomes. The 44 autosomes have 78 limbs or parts.

Members of the primate Family Cercopithecidae have the following chromosome findings. Allen's swamp monkey *(Allenopithecidae nigroviridis)* has 48 chromosomes. There are 92 parts to the 46 autosomes. *Circopitheus aethiops sabaceus* (vervet monkey, African green monkey) has 60 chromosomes. *Macaca mulatta* (Rhesus monkey) has 42 chromosomes with 80 parts to the 40 autosomes. *Papio sp* (baboon) has 42 chromosomes with 80 parts to the 40 autosomes. *Nasalis larvatus* (proboscis monkey) has a diploid chromosome number of 48.

Chromosome findings of the Family Pongidae follow. *Gorilla*

gorilla gorilla (lowland gorilla) has 48 chromosomes with probably more than 78 parts to its 46 autosomes. *Hylobates concolor leucogenys* (crested gibbon) has 52 chromosomes with 94 parts to 50 autosomes. *Pan tooglodytes* (chimpanzee) has 48 chromosomes with 80 parts to 46 autosomes. *Pongo pygmaeus* (Orangutan) has 44 chromosomes with 74 parts to its 42 autosomes.

Some New World monkeys, Family Cebidae, have the following chromosome numbers and parts. *Ateles geoffroyi* (black handed spider monkey) has 34 chromosomes with 62 parts to the 32 autosomes) and *Saimiri sciureus* (squirrel monkey) has 44 chromosomes with 72 parts to 42 autosomes.

The primate Family Lemuridae includes the lemurs. Chromosome numbers available for seven species reveal seven different numbers from 44 to 66. The family Tupaiidae includes *Tupaia glis* (tree shrew from Malaysia) which has a diploid chromosome number of 60 with 70 parts to 58 autosomes. Other shrews from Thailand and Borneo have 62 and 64 chromosomes respectively.

It should be quite clear that it is impossible to know the size, shape, and number of chromosomes of any animal based upon fossils. Without that information, descent from those animals is sheer, pure speculation and in the face of conflicting evidence. Chromosome numbers are now known to be a stabilizing influence and inhibit the pliability of species that is required in the theory of evolution. Common ancestors are impossible for different species of the same genera which have been studied.

It is noted above that none of the other primates could be related to the human, *Homo sapiens.* None have either the same chromosome number or even the same number of parts of chromosomes. It is proven that none of them could be parents to the human, and the present status is that a common ancestor is impossible. The theory of man's descent from a monkey with 48 chromosomes was made a shambles when it was found that instead of 48, man has 46. The chromosome numbers being generally similar from 42 to 50 of monkeys and large apes is consistent with the concept that similarity in body structure may at times be correlated with chromosome structure and numbers without a descent or ancestry relationship.

The classification of plants and animals is based upon characteristics which distinguish groups of various sizes. The separation from orders into families, genera, and species is based upon features common to each group which are different from other groups. The greater the apparent difference, the earlier the separation. Thus, two families have much greater visible differences than two species of the same genus. As for real ancestry or descent relationships, the classification is only speculative. In general it correlates with the theory of evolution. That implies that the great changes occurred before the minor differences between species, an apparent inconsistency.

The relationships suggested by the classification and by the theory of evolution do not correlate with chromosome findings of recent years. For a real ancestor-descent relationship, DNA of chromosomes must be transmitted. Real relationships have been disproven as to all except for members of the same species. There was previously no observable, repeatable scientific evidence that favored evolution over multiple originations, and now scientific evidence is strongly against evolution having occurred. Not only have scientists disproved their theories of origins of life by inability to produce the necessary chemicals under postulated conditions, they have disproved the theory of evolution by genetic studies.

The present status of the theory of evolution is that it is probably not true. That is the scientific status regardless of psychological, philosophical, and political barriers existing in the scientific community to admit it.

A look at plants and animals with this perspective will show how the new theory explains their existences and relationships better than the theory of evolution.

XIV. THE FIRST ANIMALS

There are today approximately 40,000 species of one-celled animals in the phylum Protozoa. There are from 15,000 to 500,000 species of nematodes or roundworms. It is proposed that those creatures originated as separate conglomerates of units of life such as DNA and RNA in situations where those chemicals were produced in abundance. By chemical reaction standards, to produce 500,000 species would require very little DNA and RNA, far less chemical material than required to make a woman's nylon stocking. By way of comparison for amount of DNA and RNA required, one cubic centimeter, one-fourth of a teaspoonful, of prostate cancer contains approximately 1,000,000,000 cells. That is about 1,000 times the number of species and animals that exist today. A single ejaculation of a normal adult male human has at least 500,000 spermatocytes in addition to fluid and nutrients. Chemical reactions are commonly characterized by varieites of resulting chemicals and side reactions which would cause variations in types of DNA and RNA required to produce the different species. Had those spermatocytes been produced chemically, they would have been capable of producing many different species.

On the other hand, to change one initial one-celled organism so that it gives rise to 40,000 different types of cells with each one needing to be isolated in order to produce a new species is an extremely formidable matter. That is just the start of what was required of that most primitive form of life. The production of 1,500,000 living species plus all of the extinct ones in such a manner strains all imagination. The formidability of the task is emphasized when it is realized that not one new species has been produced under observation even with diverse experimentation directed to the problem.

To produce single-celled germ cells in the chemical vat that could produce 40,000 types of one-celled animals would require but little more. Those germ cells would eliminate any need for evolution.

The oldest fossils of animals are not those single-celled animals. There may be a few tracks of complex worms in the precambrian system. The Cambrian had a variety of complex organisms including extinct small bivalved ostracods and a variety of snail-like molluscs. Brachiopods, a group of bivalved animals, now rare, were abundant. Approximately 75% of the fossils of the Cambrian layer were trilobites. Now extinct, most were about 3 inches in length but species varied in length from one-fourth inch to almost two feet. *Figure 13.* Such complex creatures hardly correlate with the gradualism of the theory of evolution.

Figure 13. One of many species of trilobites, complex arthropods common in the Cambrian system. The only earlier animal fossils are tracks of worms.

The first separation of living organisms is into plants and animals. A major difference is that most plants utilize carbon dioxide with oxygen as a waste product while animals utilize oxygen and excrete carbon dioxide as a waste. The structure required for chlorophyll in plants to function is relatively simple. The chlorophyll absorbs energy from the sun and polymerizes molecules of carbon dioxide to form sugars utilizing hydrogen from water and the excess oxygen is released into the atmosphere.

The hemoglobin system in animals is much more complicated. The hemoglobin absorbs oxygen in the lungs or gills and when it reaches the tissues of the body it releases the oxygen. Carbon dioxide is transported from the tissues to the gills or lungs, released, and excreted into the water or atmosphere. The system requires transport of hemoglobin back and forth between tissues low in oxygen yet high in carbon dioxide and a site where the reverse is true and oxygen can be picked up and carbon dioxide can be excreted from the organism. The hemoglobin is of no use without a complete circulatory system. The circulatory system would be of no use without hemoglobin. Construction of hemoglobin and a complete circulatory system at one time is a gigantic change in the postulated process of evolution that probably required only minor differences of sequences of nucleotides in the process of producing a batch of DNA.

It has been pointed out that there are other parts of animals that required total production to be useful. Partial or gradual changes would have been of no survival benefit or usefulness of the following: hair of mammals, feathers of birds, segmentation of arthropods and vertebrates, transformation of gill arches, teeth, shells of molluscs, poison apparatus of snakes, whale bones to enable the tail to move up and down, compound eyes which involve cornea, lens, retina, optic nerve, and receptors in the brain, and the mammalian ear with drum, oscillating bones, hearing organ, auditory nerve, and receptors in the brain.

All of the foregoing that are real strains to the theory of evolution could have been accomplished in the new theory of multiple originations by expected differences in products in chemical reactions.

XV. THE PARTS OF WHICH ANIMALS ARE MADE

It is well recognized that each of the large land bodies, the continents and islands, has its own distinct flora and fauna. Differences may be as slight as differences between species or as great as differences between families and orders.

In general plants and animals of North America resemble those of Europe and Northern Asia more than they do those of contiguous Central America and South America. Plants and animals of Central and South America resemble those of Africa more than they do those of North America. Similarly, plants and animals of Africa resemble those of South America more than they do those of Europe and Asia, except for the Asian Subcontinent. Land bridges between North America and South America and between Africa and Europe apparently have had little or no influence on the totality of specific flora and fauna of different continents.

In the theory of multiple creations, the flora and fauna of different continents are different from each other because for the most part they were separately created on each of those continents. The plants and animals of a specific continent have similarities because they originated from the same sources. They are similar to those of other continents because they resulted from similar chemicals and similar processes.

As might be expected from various chemical combinations, there is a broad mixture of resulting plants and animals. The nucleotide sequences that form the genetic code in DNA give messages as to which amino acids are to be used to make proteins. Depending upon the order of the four nucleotides in the DNA strand, the code of which amino acid is selected in sequence in the chain of a protein it is building depends upon groups of three nucleotides. For example, three deoxyadenylic acids in sequence in the DNA is the code for the amino acid phenylalanine. Larger groups of nucleotides probably determine larger parts of animals such as shape of ears, shape of nose, types of horns or antlers, types of feet and hoofs, etc., and sizes. In plants those larger groups probably determine the type of stem, shape of leaf, shape of flower, type of seed, etc.

Probably no animal has been used to demonstrate claimed evolution more than the giraffe. It is really a very poor example or fact upon which to base that theory. The animal may grow to a height of about 20 feet, while its nearest evolutionary "relative," the okapi, stands about 6 feet tall. The long neck can be used in various versions of causes of evolution. In the use-disuse version, the elongated neck resulted from stretching for higher leaves. The mutation version is that the first born with a long neck and its progeny were able to reach leaves when those with shorter necks died from starvation. The filling of a niche idea is that there was a place for a tall animal with a long neck to reach leaves of trees that would otherwise be wasted, so nature provided it a long neck to fill that niche.

The evidence that evolution occurred that produced the giraffe is extremely poor. All of the claimed fossil precursors are much smaller, and in none is the neck extremely long. The horns are very different. Those extinct claimed-to-be precursors were scattered over Europe and Asia. There are no animals in existence that show any evidence of a tendency or trend toward developing such a long neck. Inasmuch as there are no similar long-necked extinct or existent mammals on continents other than Africa, the concept of its resulting from a niche lacks merit when the same sort of niche existed on other continents. Short animals, such as antelope, camels, zebra, and okapi, have survived in Africa. There is no reason to believe that a short giraffe would not have survived or that there was one.

It is submitted that the real reason the giraffe has survived has not been its ability to eat leaves from high branches of trees. It is postulated that the giraffe has survived simply because there has been an abundance of prey for lions in Africa. Had they really needed them for food, the lions could have easily killed all the young giraffes.

The family Giraffidae has only two genera and each genus hs only one species. The term "endemic" is used for a plant or animal that is found in only one location. Both of the species in the Giraffidae family are endemic to Africa. One is the giraffe *(Giraffa camelopardalis)* and the other is the okapi *(Okapia johnstoni).* The okapi is about five feet high at the withers and has no extra elevation of the shoulders as seen in the giraffe. The diploid chromosome number of the giraffe is 30; and that of the okapi is 45. As noted above, with such numbers a common ancestor is impossible.

Little attention has been paid to the color patterns of the giraffe and of the okapi. The color pattern of the giraffe is distinctly different from that of all other hoofed animals. The color patterns vary slightly in varieties of giraffes, but all are basically medium-brown spots on a light tan background. In one subspecies, the tan forms a fine reticulated pattern. The only mammals having similar spots are leopards, and they are in different patterns. Reticulated patterns are seen in frogs and snakes, but even the most ardent evolutionists probably do not attribute the pattern to that source. Even mammals that dwell in trees do not have a similar coloration. The giraffe simply has a coloration that cannot be explained on the basis of evolution.

The color pattern of the okapi has no resemblance to that of the giraffe. Its thighs and legs are striped like those of a zebra. *Figure 14.* While the zebra lives in open areas, the okapi lives in dense forests, so the stripes cannot be attributed to camouflage for a similar environment. Contrary to what might be expected, the giraffe has a mane like that of the zebra, while the okapi does not have a mane.

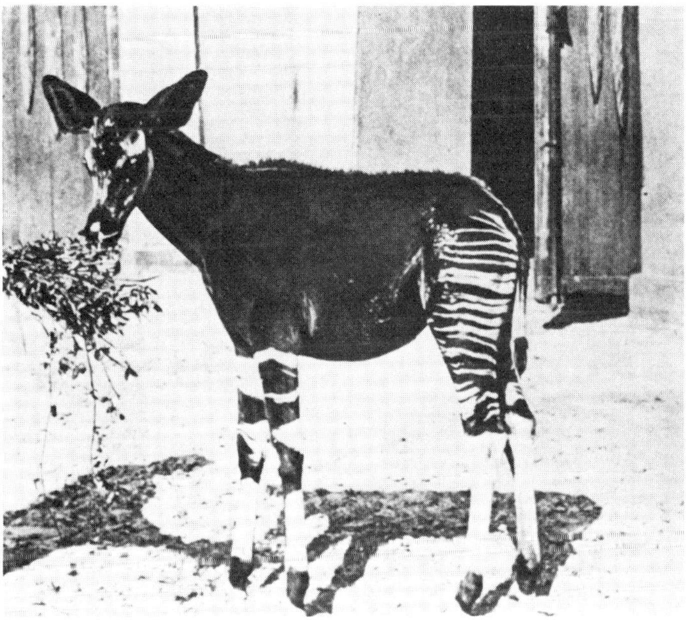

Figure 14. *Okapi johnstoni*, an animal endemic to Africa, and claimed to be the nearest living relative of the giraffe.

The giraffe and okapi are apparently unique creations resulting from a pool of chemicals of life on a specific continent. The giraffe has a unique combination of features that are both unique but similar in part to those of other animals endemic to Africa. In general, the giraffe is similar to other African herbivorous animals. It has four legs, divided hoofs, gives birth to live young, suckles them, etc. Some of its parts are completely different while other parts resemble those of other animals. The elongation of the neck, its coloration, and its horns are unique. Its horns are not horns at all. They are bony protuberances covered with hair. (Claimed fossil precursors had antlers.) Varieties or subspecies may have one or three such protuberances in addition to the larger pair. On the other hand, its long mobile lips resemble those of the camel, and its mane is similar to those of zebras and African Wild Asses.

The okapi has a head and tail that resemble those of the giraffe. Unlike the giraffe, the female okapi is hornless. The okapi has no mane, and its lips are shorter than those of the giraffe. Its hind quarters are similar to those of the camel. It is dark brown, with stripes on its legs and hind quarters resembling those of the zebra. It looks as if it was built with "spare parts" from a giraffe, a camel, and a zebra. It is exactly what one would expect to occur with some animals in the multiple originations theory. It appears to be the result of a combination of parts of DNA of animals that are not at all alike in total structure. The combinations involved cannot be explained in the theory of evolution.

Colorations of other animals of Africa also are distinctive of the continent without correlation with ancestry as postulated by the theory of evolution. Although seen on a few animals on other continents, vertical stripes on the bodies of animals seem to be characteristic of several animals of Africa and of the contiguous Asian Subcontinent. Conspicuous examples of African animals with vertical stripes are the zebra, bongo, eland, kudu, nyala, bush-buck, wildebeest or gnu, the banded mongoose, and the tabby house cat. The Bengal tiger which inhabits the contiguous Asian subcontinent is similarly striped. The tasmanian wolf is similarly striped, a color pattern which is apparently uncommon in its native Australia-Tasmania area. There are no conspicuous vertically striped animals native to North America or South America. The young Brazilian tapir has very different horizontal body stripes. The agouti is a South American rodent the size of a hare that has strikingly similar horizontal stripes. As for camouflage value, there is no evidence that stripes are an evolutionary improvement over the usual dull coloration of most animals. Both zebra and tiger might have done better with some other color arrangement.

The African lion is an animal which can be observed in zoos, and its appearance is within everyone's knowledge. The female lion is not particularly distinctive. She has a body shape and hair distribution similar to other members of the cat family including tigers, leopards, and puma. The male lion is quite different. No other cat has long hair surrounding its neck or a mane of any type. The distribution of hair is similar to such unrelated animals as the American bison and the lion marmoset, a South American monkey. The human male has long hair on the front and back of his neck but not on the sides, and the male and female horses, asses, and giraffes have long hair on the back sides of their necks.

The male lion has another unusual characteristic. It has a hard, scaly, claw-like structure on the end of his tail. The rattlesnake has modified scales at the end of its tail and American monkeys have prehensile (grasping) tails that have no hair on their underside and "fingerprints" at the end. From other peculiar features at ends of tails, a claw-like structure at the end of the tail of an animal is not an extreme variant.

If there is claim that the mane provides a sexual attraction advantage, it is based upon speculation and not experimentation. As for the claw on the end of the tail, it is inconceivable that it provides a sexual advantage, aid in defense, or aid in the hunt. If the mane and claw provide any sort of advantage, then the puzzle would be why other predatory animals don't also have them.

The theory of evolution depends upon changes that took place over hundreds of millions of years. If female and male lions resulted from the same ancestors, and their ancestors and they mated over such long periods of time and were pliable as the theory of evolution requires, it would be expected that the male and female offspring would gradually have come to look alike. The differences of the male and female African lion are unexpected in that theory.

The concept of evolution acting to fill niches requires demonstration of a niche that is related in some way to survivability. The mane and tail claws are not such niches. Further, other animals have not found such a niche to fill.

In the theory of multiple creations, the African lion poses no problem. The extreme difference in male and female is accounted for by separate origins with consequential differences of sperm and ova that had enough similarities in chromosomes to have been compatible for producing offspring. The long hair about the neck is a characteristic of nucleotide sequences that was available for forming animals that are very dissimilar, such as bison, monkeys, humans, and horses. In this instance it was mixed in which the chromosomes required to produce a cat. The unusual claw-like structure is just another variation of nucleotide sequences available for the end of a tail. The tuft of hair at the end of his tail is another example of tail genes available which is seen in such diverse animals as the rhinoceros, African wart-hog, giraffe, eland, Colobus monkeys, cattle, okapi, zebra, and elephant. It is impossible to account for the tuft of hair at the end of the tail of the male lion by the theory of evolution. The American bison has a similar tail; otherwise, it seems to be mainly an African trait.

There are many animals other than the male lion and giraffe that have unique characteristics which cannot be explained by evolution. As separate creations unique characteristics are expected to occur occasionally. A few of the many examples will be presented.

The armadillos consist of 9 genera and 20 species endemic to the southwestern part of the United States, Mexico, and Central America. Their armor-like coating is unique for mammals. There is no fossil sequence of change in skin or hair to produce its armor coat. The armadillos have no known ancestry and no close relatives. The hard outer coating has unique characteristics, but in the pool of genes it may be related to shells of tortoises, crustaceans, and insects.

The armadillos were formerly considered to be in the same family as the sloth and anteaters, also endemic to South America, but they are now considered to not be related. Chromosome studies have confirmed that conclusion. Hoffman's two-toed sloth *(Chloepus hoffmanni,* Family Bradypodiae, order Edentata, class Mammalia) has a diploid chromosome number 49. The diploid chromosome number of the soft-tailed armadillo *(Cabassous centralis,* Family Dasypodidae, order Edentata) is 62, of the hairy armadillo 60, of the nine-banded armadillo 64, and of the six-banded armadillo the number is 58. The diploid number of the collared anteater is 54. The chromosome numbers, shapes, and sizes establish that there is no connection between any of those species. There is no scientific basis for speculation that the foregoing animals could be related to a common ancestor. Armadillos are simply unique creatures for which there is no explanation in the theory of evolution.

The American pronghorn is another animal with unique characteristics and no known ancestry. Its name and classification are *Antilocapra americana,* family Antilocapridae, suborder Ruminata, order Arteriodactyla, class Mammalia. It is the only member of the family. It is endemic to the western part of the United States, and is noted as being able to run faster than any other animal in America. It prefers to roam grasslands and sage brush country in herds and avoids timberland and forests. Its horns are short, varying between about ten to fifteen inches in length, and curve inward at the tips. The horns of the male are bifurcated near the tips. They are not only unique in shape, but more significantly in their structure. They have a short bony core that is surrounded by a softer, keratinous, hollow shell. The hollow shell is shed each year as a new shell forms underneath it. The hair also has unique characteristics. It is flexible in early summer, but later in the season it grows longer, coarser, and more brittle. Then the points of those hairs rub off, leaving a thick underneath coat of fur for winter.

The structure of the horns differ from antlers of deer in that antlers are bone. The horns of true antelope are permanent coverings for a bony core. They are not shed.

There are no true antelopes in North America. The pronghorn obviously has no close relatives in North America. Its unique horns make it impossible to claim that it is a close relative of Eurasian or African antelope. Its existence cannot be explained in the theory of evolution. Its 58 chromosomes compare with 58 and 60 chromosomes of species of African antelope and 60 chromosomes of the American bison revealing again correlation of chromosomes with body structure without real relationship.

Certain deer of North America have antlers of different shapes than those of other continents. Most members of the deer family have antlers which have major stems which swing up and backward with tines that increase in number each year. The American white-tailed deer *(Odocoileus virginianus)* is found in the eastern and central parts of the United States. It has major stems which curve forward. Regular tines come off topside. The mule deer *(O. hemionus)* is found in the western part of the United States except it is not found west of the Cascade Mountains. Its antlers have no major stems, but instead each divides into two essentially equal branches the second year and then by the fourth year each of those branches form two more, making four regular points on each side. As the animal becomes older, additional tines may branch off irregularly superimposed on the basic four branches *Figure 15*.

Those two patterns of antlers on the American white-tail deer and the American mule deer are apparently unique. There is no reason to believe that the patterns are related to survivability or evolution.

Differences between flora and fauna of various continents are exaggerated in the instance of Australia. Some of those differences will be presented.

Figure 15. Wapiti (American elk) (top) with antlers having the general configuration of those of most deer. American white tail deer (middle) and American mule deer (bottom) showing the unique configurations of their antlers.

Mammals native to Australia are humans, dingo dogs, bats, rats, and its peculiar montremes (platypus and spiny anteaters) and marsupials. Bats and rats were probably immigrants. The dingo is generally considered to have been brought to Australia by the native humans. It would not be inconsistent with this theory for those two mammals to have originated in Australia with Australia still missing the bulk of placental mammals. It is difficult to believe that either the dingo or the native human came from Africa. Possible island sources for similar humans do not have dingoes.

In order to correlate body parts of Australian monotremes and marsupials, they must be considered in context with the geologic periods in which they originated. As noted above, marsupials and presumedly monotremes originated in North America during the Jurassic Period. Australia has had no significant crustal movements since that time. Its mountains correspond to the Appalachians. The forces that caused the Rockies, Andes, Alps, and Himalayas missed Australia. Australia consequently missed most or all of the volcanic activity and/or petroleum seepages that have been the probable sources of originations of placental mammals. The Australian-Tasmanian monotremes and marsupials are good evidence that there has been no evolution that resulted in new species. For the past 135 million years they have not changed significantly, if at all.

Monotremes and marsupials originated at the time many species of dinosaurs, amphibia, and reptiles originated. The same basic conditions were present for all DNA productions at the time. Body parts were similar in various creatures, with and without any logical order.

The platypus (70 chromosomes) has attracted a great deal of attention. Its total length is about 60 centimeters or 24 inches. It is mainly an aquatic animal, living in lakes and streams with its home in a burrow with the entrance below the water line. Its diet consists mainly of crustaceans, mollusks, tadpoles, and worms, all of which were present during the Jurassic Period. It has a broad snout which looks like the bill of a duck. Its eyes are small and it has no external ears or apparent neck. Its body is covered with dense, beaver-like fur; and its tail is broad, grossly resembling that of a beaver. It has a cloaca; that is, its urine and feces are excreted through an opening in common with that of the genital ducts. The female lays eggs. The young are fed by her milk which exudes from pores, there being no definite teats. Its feet are equipped with strong claws, but its toes are webbed resembling those of a duck. The male has spurs grossly resembling those of a fighting cock. Each spur is connected to a gland in each thigh which makes the stab of the spur poisonous. The poison causes pain and swelling. The purpose of the poisonous spur is unknown. It obviously has nothing to do with obtaining food due to the type of its diet; and it has nothing to do with defense, since the platypus has no enemies.

From an evolutionary viewpoint the platypus is unexplainable. It is unique in totality and combines an unusual array of parts. As a separate origination from a pool of Australian DNA in the Jurassic Period it is not an unexpected combination of parts. Its snout resembles that of a duck, but in general shape it also resembles the shapes of snouts of duckbill dinosaurs. Its feet resemble those of a duck, but they more closely resemble those of its contemporaneous amphibians and crocodiles. Most dinosaurs also laid eggs. Australia has more than its share of poisonous animals, including snakes, spiders, wasps, fish and octopus. The poisonous nature of the platypus is not unusual with the DNA of Australia.

Its mammary glands, internal heat regulation, and fur were new original parts.

The echidna, or spiny anteater, (63 chromosomes) is similar in some respects to the platypus but with some great differences. Species vary from about 18 inches to 30 inches in length. It has a long, narrow, beak-like snout with nostrils and a small mouth at the end, no teeth, and no external ear. The tail is conical and short. Its back is armed with spines which protect it when it rolls up like a hedgehog. The male has a spur similar to the platypus. It too has a cloaca, and the female lays an egg which she places in a pouch where the young also suckles after birth. It obviously presents a combinatiton of features which cannot be explained by evolution.

The main distinguishing features of marsupials is the development of the offspring. The embryo of any other type of mammal, excluding monotremes, becomes attached by blood vessels to a placenta in which the blood vessels of the embryo intermingle closely with the blood vessels of the uterus of the mother. The embryo obtains its nutrition and oxygen by diffusion between those blood vessels. Marsupials do not have a placenta and the fetuses are born much earlier in their stage of development than are other mammals. Much of the development, growth, and maturation of the fetus which occurs in the uterus of other mammals occurs in the pouch of marsupials. The newborn are tiny; twenty newborn oppossums are approximately a teaspoonful. As soon as the young of a marsupial is born it crawls into the pouch where it attaches to a teat. Milk is squeezed into it by special muscles of the mother.

Marsupials are also characterized by teeth and feet different from other mammals.

In Australia and on the neighboring island of Tasmania, animals that resemble in appearance mice (opossum-mice, honey-mouse, pouched mice), guinea pigs (wombats), squirrels including flying squirrels (phalangers), rats (rat-kangaroos, musk rat-kangaroo, rat bandicoots), lemurs *(Pseudochirus lemuroides),* anteaters (banded anteater), rabbits (rabbit bandicoots), badgers (Tasmanian devils), cats (Australian native cats), dogs (Tasmanian wolf), and bear (Koala) are all marsupials. In the theory of evolution the existences of the foregoing animals are explained as resulting from evolution from some primitive immigrant opossum-like animal. Mechanisms relate to filling of niches in the ecology whether by changes in genes or just changes in body form.

The koala feeds exclusively upon leaves of only certain species of eucalyptus trees, also endemic to Australia. One surely wonders how important such a niche was. That could hardly have stimulated evolution to develop such a strange little animal. It certainly is no substitute for the niche filled by bears. If there was a niche that needed to be filled by a mammal to eat leaves off a specific species of eucalyptus trees, one wonders why there are not tree-climbing mammals to eat the leaves of other eucalyptus trees, or maples, elms, pines, etc.

One might make an analogy of the Giant Panda which is endemic to high mountains near the Tibetan border. It is a similarly strange creature that cannot be considered to be the result of isolation on an island the size of Australia, a frequently used explanation of Australian flora and fauna. It prefers a diet of bamboo shoots which it is able to grasp and break. It is believed by evolutionists to be a modified raccoon. Obviously, there are a few missing links.

The Tasmanian wolf stands about one and a half feet in height. The Tasmanian devil is smaller, and the native cats are still smaller. The head and body of the largest of the native cats measures about two feet in length. Those carnivores kill and eat birds and small animals including wallabies, but they hardly fill niches of tigers, lions, and real wolves. Most remarkable, their pouches open backwards. Instead of opening towards the head like those of a kangaroo, they open towards the tail. The difference would seem to be impossible to explain by a process of evolution.

The kangaroo has the general body shape of certain dinosaurs. Its head and color resemble those of American mule deer. Its tail resembles those of dinosaurs and of no modern animals. It apparently originated with DNA that was produced under conditions that might have produced a dinosaur but added mammary glands, pouch, internal heat regulation, different feet, different teeth, and hair.

In this new theory of multiple originations, each species of marsupials, like other animals, originated or was created as a distinct separate species. The variety of closely related species is the result of differences in chemical reactions or spectrum of products in the chemical processes involved in producing DNA at the same time or under very similar conditions. Animals resembling mice, rats, dogs, cats, deer, etc., associated with marsupial reproductive systems, feet, and teeth are not unexpected in the new theory. Instead of being examples of evolution, those varieties are just the opposite. They have probably not changed significantly since they first originated at about the time of the Jurassic Period.

As explained above, it is postulated that the main reason no (or few) placental mammals originated in Australia is because of Australia's unique geologic history. Various islands, Australia, and other continents provide an interesting analogy to the general history of the planet. Young volcanic islands such as the Hawaiian Islands and the Galapagos Islands are similar to the continents when they were first formed, with the exception of atmosphere differences. The developmental stage of Australia is the same as that of North America at the time the Appalachian Mountains were formed. Other continents have changed and added new and larger mountains and new and different seepages of petroleum. The originations of placental mammals correlates with those changes. This establishes that originations of new species at various times has probably been due to new volcanoes and petroleum seepages that occurred with crustal movements separating geologic periods. Juvenile volcanoes were generally frequent and widespread so that as the environment changed that would allow different organisms to live, new species originated and lived that previously would have died.

XVI. POISONOUS SNAKES AND LIZARDS

The flora and fauna differences of each continent includes snakes and lizards. They present some unusual differences which are of interest in the problem of evolution versus multiple originations.

There are 3000 species of lizards. Only two are venomous, and both are endemic to the southwestern part of the United States and Mexico. They are the Gila monster and the Mexican beaded lizard. There are two subspecies of the Gila monster and three subspecies of the Mexican beaded lizard. The fang is a specialized tooth with grooves and cutting ridges. The Gila monster may have 35 such teeth. The grooves allow for the penetration of the toxin which comes from glands that are located to the rear in the lower jaw.

To the contrary, glands that produce venom in snakes are above the upper jaw. Of 2,700 species of snakes, 412 are venomous. Evolution cannot explain the great difference in percentages of lizards and snakes that are venomous, the reason that the only venomous lizards are found in North America, and the difference in location of the venom gland in lizards and snakes. In that theory the lizards and snakes are closely related.

There are eight to ten families and 2,700 species of snakes. Of those, three families contain venomous species. The family Colubridae which has 1,400 species has only two which are venomous, the boomslang and the vine snake. Both are endemic to Africa and both have small fixed rear fangs. Those fangs are grooved and not tubed. The venom is a blood coagulant.

The family Elapidae consists of 230 species all of which are venomous. It includes the cobras and sea snakes. As noted above, Australia has more than its share. In all of Asia there are only 6 genera of elapid snakes, while there are 23 genera in Australia, New Guinea, and the Solomons.

The sea snakes demonstrate interesting characteristics. They are all venomous. The fangs are fixed enclosed tubes forward in the upper jaw. The toxin paralyzes the prey. Those snakes are great swimmers, yet they are confined to the Pacific and Indian Oceans. They are primarily in the Southwest Pacific. One species, *P. platurus,* reaches Central and South America and has been reported captured in Posieta Bay, Siberia. Sea snakes are mainly confined to the sea, but one species lives in fresh water on Luzon Island (Philippines) and one lives in a brackish lake in the Solomon Islands. Most of the sea snakes are found in the India and Australia areas. Their tails are flattened vertically so that although they swim well, they move poorly on land. Some lay eggs on land and some bear their young alive.

Snakes are not uncommon reptiles and many are known to inhabit and swim well in fresh water. Snakes in the same family as the sea snakes, the Elapidae, are numerous in the United States, South America, and Africa as well as in Asia. Surely many have been carried to sea in such rivers as the Mississippi, Amazon, and Congo. In the evolutionary scale they are ancient. Given the very successful adaptation to oceans in the Southwest Pacific, in the theory of evolution they would be expected to also inhabit the Atlantic Ocean either by migration or by evolution from American and African snakes. Prediction is impossible based upon the theory of evolution.

It is submitted that the sea snakes were probably constructed initially in such a manner that they could thrive well in sea water, not on land, and not well in fresh water. Snakes that were not so constructed initially never took to the sea. The sea snakes were apparently the result of variation of chemicals that produced other elapids in the region of the Southwest Pacific. It would be inconsistent to claim that they evolved from land snakes in the Southwest Pacific with no evidence of similar evolution in other parts of the world inhabited by similar land snakes.

There are three genera of Elapidae in the New World consisting of 60 species. All are endemic to North or South America. It is inconceivable that they share common ancestors with Asian snakes without some of the species being the same on both continents. Europe has no Elapidae.

Europe's poisonous snakes are all of the Viperidae family. Australia's poisonous snakes are all Elapidae. Probably the most interesting snake family is the Crotalidae, the family of pit vipers. They are found in North America, South America, and Asia.

The pit vipers have rotating fangs similar to the Viperidae family found in Eurasia and Africa. They are long hollow tubular fangs located in the forward part of the upper jaw. They fold back out of the way until the snake prepares to strike at which time they swing forward so that the fang is more or less perpendicular to the upper jaw.

Another characteristic that distinguishes pit vipers from other poisonous snakes and all other animals is its extra sense organ. Between the nostril and the eye it has a small, depressed, "pit." A membrane separates the outer pit from a space or tube that is next to the eye socket. With that organ the snake can sense the body heat of its prey up to a distance of about one foot.

Rattlesnakes are pit vipers which are endemic to the Western

Hemisphere, and most are located in the southwestern part of the United States and Mexico. The rattle is unique and consists of various numbers of jointed "buttons" which form from sections of skin which are not shed when the rest of the skin is periodically shed. The fangs of the rattlesnake are fragile. As one of the two is broken off there are up to six fangs in various stages of development ready to grow out. Their venom is toxic to both the nervous system and blood. Venom of other pit vipers is mainly toxic to the blood.

From the foregoing, pit vipers are different from other poisonous snakes by type of fang, type of poison, an additional sense organs, and rattles on some. Rattlesnakes, found only in the New World, differ from pit vipers of the Old World by type of venom and presence of rattles. A real connection is questioned.

The West Indies are located near lands on three sides that have many poisonous snakes. Only three of the islands, Trinidad, Tobago, and Martinique, have any venomous snakes. Also, as noted, the various continents have only certain families of the venomous snakes and not others. Those observations and the history of sea snakes suggests limited migration habits of snakes.

Venomous snakes present problems to the theory of evolution. The fixed hollow tube fang and venom system is a complex system. No other animals have similar structures. If it were due to evolution, some cats, dogs, lizards, alligators, fish, birds, or other animals would be expected to have such devices. Some of those animals have large "canine" teeth, "fangs" of a type, tusks, etc., without their being hollow and without having associated venom organs. From that viewpoint, the rear-fanged snakes are even more unusual. The fang of pit vipers that pivots on its base as if there were a hinge there enabling it to fold back out of the way is really a handy piece of equipment. It enables the animal to have a much longer weapon without its being in the way. There is nothing resembling it in any other animal. It must be a mystery in the theory of evolution.

The heat-sensing organ of pit vipers is an additional sensory organ. In addition to seeing, hearing, smelling, tasting, and feeling, the pit viper can detect heat of its prey or enemy. Any animal would be better equipped with such an organ. No explanation is apparent in the theory of evolution for its being found in only a few types of snakes.

The pupils of the eyes of some snakes have the shape of vertical slits when constricted similar to those of the domestic house cat. Other snakes present round pupils when constricted. It is submitted that without good solid evidence, it is unacceptable to assume any relationship between animals that have different types of eyes, including shapes of pupils. Rattlesnakes and coral snakes endemic to the American continents have different types of fangs, venom, sensory organs, tail structure, and shapes of pupils. It is imaginative to consider that they have a common ancestor. It is fanciful to consider that they are the result of evolution parallel to evolution in Asia.

Snakes and lizards, and particularly the venomous ones, demonstrate examples of organisms whose existences are better explained as being the results of multiple originations or creations than as due to evolution.

XVII FLIGHTLESS BIRDS

Many birds are either poor fliers or unable to fly. The common cootie is an example of a poor flier with which duck hunters are acquainted. A few birds that are unable to fly are conspicuous. Penguins are common examples used in presenting the theory of evolution. It is commonly claimed that because of disuse the wings of the penguins were changed to flippers. Modifications of the theory of evolution claim the change was due to mutations to the advantage of the penguins that made them more fit to survive. As with similar examples, there is no evidence that the penguin was ever a flying bird. There is no evidence that the penguin would not have survived as a flying bird, or that it would not be happier if it could fly.

Cassowaries are flightless birds endemic to Australia, New Guinea, and neighboring islands. The largest, *Casuaris casuarius*, stands about five feet tall (1.5 meters). It is slightly smaller than the emu, which in turn is slightly smaller than the ostrich. The head is covered with a bony helmet, or casque, which is unique in the modern world. It would not have appeared unusual in company with the crested dinosaur, Corythosaurus. Instead of the usual four toes of birds, it has only three. The inner toe is equipped with an unusually long claw that serves as a dagger when the bird kicks.

Ordinary feathers are constructed with a central shaft and paired branches or barbs forming a flattened slightly curved surface, the vane. The barbs have small branches, or barbules, that interlock adjacent barbs giving firmness to the flat vane shape of the feather. The feathers of the cassowary lack the barbules and feel soft and hairlike as a result. Five long quills that extend from the end of each wing have no barbs.

Kiwis are flightless birds about the size of chickens endemic to New Zealand. An egg may weigh up to 450 grams (one pound), the largest relative to the size of the bird. Their feathers are soft and hair-like. It is almost wingless, its wings being small, vestigial structures hidden under the feathers. Their nostrils open at the tip of the long flexible bill rather than at the base which is usual for birds.

It is noted above that the echidna, or spiny anteater, found in nearby Australia and Tasmania also has its nostrils located in the tip of its long beak-like snout. The unusual "hair" spines of the echnidna, the unusual feathers of the kiwi, and the nostrils at the tip of their beaks suggest a possible relationship between the kiwi bird and the echidna. That relationship is unexplainable in the theory of evolution and fits the pattern of multiple originations.

The emu is a flightless bird found only in Australia. It may be as tall as 6 feet (1.8 meters) and may weigh up to 120 pounds (55 kilograms). Females are slightly larger than males. Like the cassowaries, the emu has only three toes. It has a beak flattened from top to bottom, but not widened like a duck bill.

The ostrich is the largest of the living flightless birds. It is endemic to Africa. The male is larger than the female and may stand nearly eight feet (2.5 meters) in height and weigh 345 pounds (155 kilograms). Instead of four toes as the usual bird and kiwi have on each foot, or three as the emu, cassowary, and rhea have, the ostrich has only two toes on each foot, with one considerably larger than the other. Its feathers, like those of the kiwi, have no aftershafts.

The rhea is a large flightless bird endemic to South America. It is shorter and lighter in build than the emu, the ostrich, and the cassowary. It may stand four feet (120 centimeters) tall. It has three toes on each foot.

In the theory of evolution, the similarity of cassowaries, rheas, emus, and ostriches has suggested a common origin. This involves flightless birds on Australia, South America, and Africa. All of those large land bodies are separated from each other by large bodies of water. The theory of evolution of those birds from a common source may require land bridges for which there is no evidence.

The theory of evolution requires parts to be usable, otherwise they are of no benefit to survival. It postulates that those birds could formerly fly in spite of the first fossil birds being non-flying. If that postulate were true, one would expect at least one of them or some similar bird to be able to fly. The inability of the rhea, ostrich, and emu to fly can hardly be attributed to smallness of the land masses they inhabit.

In this theory of multiple originations, parts are not guaranteed to work. The bodies of those birds as well as their wings are not structurally built to fly. In addition, their feathers are not constructed for flight. It is postulated that when DNA was produced, usable combinations resulted in many cases. Some probably didn't work at all and the result was a creature that couldn't live. Other combinations probably caused great but not lethal handicaps. Animals adapt in this theory not by changing the anatomic structure of their offspring, but by doing the best possible with what they have. That is the type of adaptation we observe and can observe repeatedly. That is the situation of those flightless birds. They don't chase insects in the sky. If they wasted their time doing that, they would die. So the cassowary has found fruit and small animals to eat, ostriches live on plants and animals, and penguins catch fish.

There are reasons to suspect that low oxygen concentration had something to do with originations of flightless birds. The earliest birds were probably flightless, and the fossil sequence seems to have been due to an increasing oxygen supply. Furthermore, it has been observed in Colorado that at elevations of over 10,000 feet where oxygen is deficient, chickens usually do not hatch. When the oxygen was deficient, it may have been that only embryos of flightless birds lived.

A common factor is that those large flightless birds are endemic to the Southern Hemisphere. If they were migratory as postulated by proponents of the theory of evolution, one would have expected them to inhabit North America, Europe, and Asia. In Africa, South America, and Australia, there has apparently been an abundance of food for both herbivorous and carnivorous animals so that even misfits and relatively helpless have been able to survive. That is not equated with evolution, but only with why they are not extinct. Had they been able to produce offspring more fit to survive, they would have done so.

XVIII. ENDEMICS

In this theory of multiple originations and creations of living organisms, the flora and fauna of different land masses have a general similarity. That is because the chemicals of petroleum are essentially the same everywhere and other conditions required to produce the chemicals of life are similar regardless of location. Occasionally two separate creations on different continents are almost indistinguishable. On the other hand, because of subtle differences in petroleums and differences in conditions in which the chemicals of life were produced, nearly all organisms on one continent or island are different from those on other continents and islands. Some of the continents and islands have entire families and orders that other land bodies do not have, and some continents have conspicuous absences of entire families and orders. Because of the nature of multiple creations, animals may exist that are very different from any other animals.

As noted above, a plant or animal is said to be endemic to an area if that is the only place it is found. The size of the area of endemicity varies. It may involve only a small part of a continent, an island, a group of islands, an entire continent, or a hemisphere.

It has been observed that each continent and large island has its own distinct flora and fauna. Nearly any native animal that comes to mind is endemic. For example, the mule deer, white tail deer, pronghorn, Wapiti (elk), American bison, coyote, jack rabbit, sage grouse, turkey, rattlesnake, cotton mouth snake, alligator, puma, lynx, beaver, and grizzly bear are conspicuous animals that are endemic to North America. African animals with which we are all acquainted are not found on other continents. The same phenomenon occurs with plants, although most of us are not as aware of it. Sage brush, cactus, giant sequoia, redwoods, Monterey cypress, bristlecone pine, and Joshua tree are a few examples of plants endemic to the western part of the United States.

The theory of evolution accounts for differences in flora and fauna of various continents and islands as being due to isolation and change. It is completely speculative, as essential postulated changes have not been observed. The multiple creations theory merely claims that given the fact that at least one living organism originated, that more than one have similarly originated. Given the multiplicity of organisms and observation that species reproduce the same species only, the logical conclusion is that diversity of organisms is probably due to differences at the time they originated.

Primordial conditions of the Earth cannot be reproduced; but fortunately all land masses on Earth are not the same age, and all have not been subjected to the same forces or changes in their geologic histories. When Earth was first formed there were probably no sediments. There was probably much "outgassing" from the interior of the earth through volcanic vents, along with flows of magma, or lava. The gases included water as steam, which precipitated and formed oceans. The land was mainly of volcanic origin. The atmosphere was mainly that produced by volcanoes, mostly methane, ammonia, nitrogen, carbon dioxide, and water.

Young volcanic islands probably resemble the conditions of continents on the primordial Earth except for atmosphere differences and particularly an increase in oxygen. Conditions unfortunately have not been controlled and the role of invaders is subject to differing opinions.

The Hawaiian Islands are relatively young volcanic islands that are quite isolated from other land masses. An examination of their flora and fauna can supply information pertinent to the problem of evolution versus multiple originations. Of 85 species of land birds on the islands, 70 are endemics. It has 3,722 endemic species of insects and only 250 that are found elsewhere. Of about 2,000 species of seed plants, 1,633 are endemic. There are 1,061 endemic species of land snails and 25 that are found elsewhere. That is a total of 6,057 species of which 5,175, or 85% are endemics.

The basic question is whether some or all of the 6,000 species of flora and fauna of the Hawaiian Islands originated there as new creations of life or whether species of plants and animals arrived there from all parts of the Earth and then 85 percent of them changed to have unique characteristiccs.

First, it must be recognized that if 4000 insects originated on the Hawaiian Islands and 4000 insects originated on some island near Viet Nam that some of them would probably appear to be of the same species. All members of any species are not identical, and a range of variation is allowed. Only 250 insects being similar is not inconsistent with separate originations of them along with the 3,722 endemic varieties. A similar situation exists with plants.

It is generally believed that birds arrived on the islands having eaten seeds which were planted with excrement. Ocean currents and winds brought birds, insects, and seeds. From those sources, the plants and animals changed to constitute the present flora and fauna. It is not clear how the snails are postulated to have arrived.

The Hawaiian Islands are north of the equator approximately 3000 miles from Baja California. Predominant ocean currents are from that direction, from Mexico to Hawaii. The flora of Hawaii includes plants similar to those of Central America and as far away as Madagascar and Africa.

Plants endemic to the Hawaiian Islands lend support to the new theory of multiple originations. Some of them are "archaic" types which reproduce differently in alternate generations. One generation reproduces by spores and the next generation reproduces by male sex organs with motile sperm and female sex organs with ova. These plants are similar to plants that flourished 300 million years ago during Carboiniferous periods and formed coal, and have since been extinct.

The Hawaiian "uluhe" ferns are primitive ferns similar to ancient extinct varieties. The spores are exposed on the underside of fronds while in ferns that appeared later the spores are covered. *Dicranopteris emerginata* is one of those primitive ferns endemic to only the islands of Oahu and Hawaii. If it is maintained that those plants grew from spores carried by the wind 300 million years ago to some islands that sunk but were precursors to the Hawaiian Islands, the obvious problem is why they are restricted to only two islands. If spores were carried more than 3000 miles it seems inconceivable that they would not have spread to other islands. Similar phenomena of endemics of specific islands in the Hawaiian group is the rule and not the exception, consistent with multiple originations on those islands.

There are three endemic tree ferns which are also similar to fossils in coal. Other endemic ferns are similar to those of the Old World and others are similar to some native to South America.

The coconut of Hawaii is endemic. It is not as closely related to the New World type as to one of southern Africa. That would require at least 10,000 miles of ocean voyage to reach Hawaii. Folklore relates that Hawaiians took the coconut to Madagascar.

The wild Hawaiian pineapple is endemic. The domesticated variety is a native of Brazil. A spread by wind or sea is quite improbable.

There are only three orchids, all of which are rare, endemic to the Hawaiian Islands. Orchid seeds are as light as dust. Why Hawaii has only three types when there are numerous varieites in other tropical areas is a mystery to those who attribute their presence in Hawaii as due to invasion from some other land. The endemic Hawaiian poppy has its closest relative in Florida and a more distant relative in Colorado. There are numerous endemic herbs and shrubs including some that grow on barren lava slopes and within volcanic craters. It also has a large raspberry that is too bitter to eat, and a relative of the apple tree.

As noted above, Hawaii has 1,061 endemic species of land snails. They include tree snails that have brightly colored shells which are spiraled in helical cone shapes about one inch in height. They occupy bushes and trees throughout the islands with each isolated valley having snails of its distinctive types.

The birds of the Hawaiian Islands are also of great interest. It should be noted at the outset that birds of one species do not normally breed with those of other species when in the wild. There are 38 species and 38 subspecies endemic to the Hawaiian Islands. They involve 16 endemic genera and represent 11 families. Three of those bird families have representatives in the Old World and most of the others have representations in North America and Asia. In other words, the birds as classified into families represent families that are scattered all over the globe. With the exceptions of certain ducks and shorebirds, the Hawaiian Islands are not even near any migratory routes.

The Hawaiian Islands have as endemics a large number of species of small song birds, 4 species of rails, a crow, a hawk, an owl, a small, teal-sized, insect-eating duck, another large duck, and a small goose. The endemic birds tend to stay isolated on certain islands while introduced birds become widespread. That is noted as evidence that the endemics were not invaders, but originated on the islands.

The small Laysan duck lives only on Laysan Island and Lisianski Island which is 140 miles northwest of Laysan. The small millerbird, which is less than six inches in length, is found only on isolated Laysan and Nihoa Islands, each having a separate species. The single species of crow, hawk, and goose are found only on the island of Hawaii The goose is found at elevations between 5000 and 8000 feet where it nests and feeds on vegetation on lava flows. It returns year after year to the same area to nest. Pinpointing its origin and site of initial birth to a specific volcano, those pen reared on Mauna Loa have returned to Pohakuloa to nest.

The songbirds are of particular interest. The family Corvidae (crows, magpies, jays) is represented by the Hawaiian crow. The Hawaiian thrush represents the family Turdidae (thrushes, robins, bluebirds, solitaires). The family Sylviidae (Old World warblers, New World gnatcatchers) is represented by two species of the millerbird. The Old World flycatchers, family Muscicapidae, is represented by an endemic genus, Chasiempis. Distinct varieties are found on Oahu, Hawaii, and Kauai only. The honey-eaters, family Meliphagidae, has two endemic genera with five species located on four islands.

The birds that have received the most attention and are the most distinctive on the islands are the Hawaiian honeycreepers, the family Drepanididae. The entire family of 9 genera, 22 species and 24 subspecies is endemic. They are of unknown origin. These small birds are mostly about the size of canaries or sparrows. They vary greatly in color with all sorts of combinations of black, browns, yellows, and reds. Their most unusual characteristic is their assortment of beaks. They include beaks like those of ordinary canaries, some like those of parrots, some that are long, thin, and curved, some that are heavy like a puffin's being almost the size of the rest of the head, and other varieties. They resemble in shapes and varieties the beaks of Darwin's finches endemic to the Galapagos Islands.

As noted above, Hawaii's native animals are insects, snails and birds. Those animals must be considered in context with the types of animals that are missing. Instead of being evidence of evolution, they present conflicts with the theory of evolution. First, from a gross viewpoint, there are birds without preceding reptiles. There are no native walking fish, lizards, flying lizards, snakes, marsupials, mice, squirrels, rats, cats, dogs, sheep, goats, deer, etc. There are obviously gross niches for such animals, and varieties introduced by humans have thrived.

If the birds flew in pairs more than 2000 miles from America and more than 10,000 miles from Africa to arrive at the Hawaiian Islands, they would be expected to fly between the islands and inhabit all of them instead of being endemic to specific islands. If those little song birds could have changed their beaks, according to the theory of evolution they could have changed their bodies. There was plenty of area and food for a real parrot, a real woodpecker, and humingbirds. (It might be noted that humingbirds are limited to the Americas without fossil ancestors.) The crow did not evolve to form related jays or magpies. The thrush did not generate robins. There are no ravens and no vultures. The owl and hawk, being limited to one species each, showed no tendency to evolve to fill niches. There was certainly room for more than two species of ducks on all of those islands, and particularly when one thrives on insects on only two small islands. They showed no ability to evolve to different species.

The Hawaiian Islands have plants with characteristics similar to plants which occurred in early geologic periods on continents.

That is consistent with their originating under conditions resembling early periods in geologic history of the continents.

Snails that were created in Hawaii are similar to snails created in primordial geologic eras elsewhere. In this new theory they were able to thrive on land in Hawaii probably because they originated there when there was adequate oxygen in the atmosphere. They may have originated in other periods in other places when oxygen was inadequate for them to survive on land.

With adequate oxygen in the air and water and plentiful nutrients, birds were created and able to thrive. Reptiles were probably never created there. That no reptile eggs or adults arrived at the Hawaiian Islands from other places is evidence that creatures of Hawaii originated there and are not invaders.

The honeycreepers are examples of a general type of small song bird that was produced in abundance in variety and particularly with a variety of the part of DNA that determines the shape of the beak. They adapted to use those beaks to their best advantage. In this new theory of multiple originations and creations, contrary to the theory of evolution, offspring were not born with beaks different than those of their parents because there was some kind of food available that the beaks of the parents could not reach.

The Galapagos Islands are similar to the Hawaiian Islands geologically and in size. The Galapagos are farther south, being on the equator, and are nearer land, being about 600 miles west of South America. There are four major islands and 11 small ones. They are young, being about one to two million years old. To put this time in perspective, it corresponds roughly to the Pleistocene epoch. It is claimed that almost all modern mammal species on the continents have evolved from small similar creatures during that time period, and during a very much shorter period of time man is claimed to have evolved from monkeys or apes.

The Galapagos Islands have some small insects, giant tortoises, a few snakes, several small lizards, two species of land iguana, the only known marine iguana, and 89 species of birds. It is noted that there are no amphibia, which evolution-wise were precursors of reptiles. The reptiles show no evidence of changes toward mammals, and there are no creatures intermediate between reptiles and birds. There are no native mammals other than bats, rats, and seals, which are considered most likely to be invaders.

There are 15 subspecies of its famous giant land tortoise. Each of ten islands has its own subspecies. The largest island, Albemarle (75 miles long), has 5 subspecies each confined to the neighborhood of one of the 5 major volcanoes on the island.

There is only one species of small lava lizard on a given island, with a total of seven species. One species of land iguana is found only on Barrington Island, and the other species is found on five islands. Species of the family Iguanidae are found in North America, South America, the West Indies and the Galapagos Islands. Outside the Western Hemisphere, one genus is found on Fiji and Tonga, and two genera on Madagascar. It is submitted that the distribution is far too spotty and distant to support any claim of spread of those land dwellers or their eggs by ocean currents. The only marine species is endemic to the Galapagos Islands. It stays close to land, and has rarely been seen as far as one kilometer from shore. They occur on all of the islands but vary in color, size, and behavior from island to island. They differ from the land species in having partially webbed toes and tails flattened to be vertical.

The Galapagos Islands have an endemic duck, hawk, penguin, dove, gull, flamingo, and two owls. They have seven species of mocking birds. No island has more than one species. One island has seven subspecies of one species. The special endemic finches are the most common birds. They are all dull in color. Six species that feed on the ground have long curved beaks. Three tree finches have parrot-like beaks. The warbler finch has a slender beak. A separate genus of vegetarian tree finch is larger. There are two species that feed on insects from tree trunks like woodpeckers. Unlike woodpeckers, they do not have long tongues in spite of a need for such. Because of that handicap, they use twigs and cactus spines as tools to pry insects out of crevices. Most of the islands have more than one species. Only one finch is confined to a single island, and one is confined to two islands.

Forty percent of the plants on the Galapagos are endemics, and similar to Hawaii, its forests are largely of plants like those of fossils.

The patterns of flora and fauna of the Hawaiian Islands and the Galapagos Islands are quite similar. They correspond to originations of living organisms correlated not only with the groups of islands but with individual islands. They correlate not only with individual islands, but even with separate volcanoes. They are evidence of no evolution in spite of great niches in the ecology that are not filled. They are evidence of plants and animals originating on those islands.

It is submitted that the new theory of multiple originations and creations provides a better explanation for the unusual flora and fauna of the Hawaiian Islands and the Galapagos Islands than explanations provided by the theory of evolution. For the most part the flora and fauna of those islands provide examples of originations of life correlated with their recent geologic history being similar to the primordial geologic history of the continents. Birds could be added because of more oxygen in the atmosphere than when Earth was young.

XIX. THE MAIN LINE

Most criticism of the theory of evolution focuses on extreme phenomena. The main problem with the theory is not the unusual, the anomaly, or the exception that it cannot explain or for which the explanation is obviously fanciful. The major problem is the main line of speculation which is inconsistent with facts, inconsistent in logic, and has no basis in scientific observation.

A basic difference in the theory of evolution and the theory of multiple originations herein is the type of cells first produced. The theory of evolution starts with cells programed only to repeat themselves. In this theory of multiple originations, germ cells originated that could produce all of the kinds of cells that make up complex plants and animals. It is like the difference between DNA of bacteria and DNA of an ovum or spermatocyte. One produces more of the same bacteria, the other can produce a human. The theory of evolution is like trying to produce babies by joining a cell from the finger of a man with a cell from the finger of a woman. Cloning of complete individuals from a body cell may be appealing to the imagination in science fiction; but from observed facts and scientific principles as known today, it has probably not actually occurred. Those body cells are programed to reproduce only their own kind of cells. Similarly, it is hypothetically possible that the offspring of a one-celled animal could be an animal of a different type than its parent; but from observed facts and real scientific principles, it has probably not occurred. In this new theory, germ cells originated at various times and places and produced a variety of plants and animals.

In the theory of evolution, simple amino acids and other hydrocarbons formed in the atmosphere and gathered together in globs of unorganized material. After hundreds of millions of years in some mysterious manner it finally became a bacteria, primarily DNA. After millions of years of bacteria reproducing the same bacteria, the offspring became a very organized one-celled organism. That cell had DNA, RNA, proteins and chemicals that could perform very complicated functions of absorbing nutrients, burning those nutrients, excreting wastes, growing, and reproducing itself.

When living organisms had arrived at the one-celled protozoa stage they had made tremendous strides in evolution. There are about 40,000 species of protozoa alive today. A teaspoonful, about four milliliters, of those protozoa would be about a million of them. An amoeba is a protozoa species that has a very mobile soft shape that enables it to ingest food by surrounding it. Other species are much more complex and have very definite external shapes, food and excretory tracts, and hair-like cilia covering the outer surface and tails that enable them to move.

Those protozoa were programed to reproduce only their own kind by the numbers, sizes, and shapes of their chromosomes. In spite of such fixed capabilities, evolutionists claim that after millions of generations of reproducing according to the genetic code, they produced worms. Other generations grew legs and wings and developed brains, eyes, stingers, egg laying and other reproductive apparatus, and became insects. Some became fish with gills, eyes, brains, fins, etc. After millions of generations of fish reproducing with eggs as fish of a specific species, some of those fish started moving over land. After millions of generations of reproducing fish that walked with their fins over land, the offspring were born not with skeletons of fish but with skeletons of lizard-like animals with legs and all the necessary parts to enable them to move about better on land than they could with fins and skeletons of fish. After millions of generations of lizards reproduced as lizards, some were jumping out of trees. After millions of generations of lizards giving birth to lizards that jumped out of trees, some gave birth to lizards with wings. Those lizards gave birth to more lizards with wings which didn't work very well. Some such lizards gave birth to lizards with feathers and wings, a more satisfactory arrangement. After millions of generations of following the genetic code of producing the same kind of species, some such lizards gave birth to real, warm-blooded birds that could sing, fly, and do everything else birds do so well. Some of the lizards gave birth through similar sequences to shrews, little animals that eat insects. After millions of generations of shrews giving birth to shrews in accord with the limitations of their DNA, some gave birth to lemurs, small primates. Lemurs gave birth to monkeys, and monkeys gave birth to apes, and apes gave birth to humans.

The foregoing is not meant to be a test of gullibility. It is a very, very, simplified version similar to what was required in the theory of evolution. To obtain a better idea of the changes that were required in offspring of that initial one-celled animal, the following list of phyla of presently living animals after protozoa is presented.

1. Mesozoa (mesozoans – small worm-like parasites of worms)
2. Cndaria or Coelenterata (cnidarious)
3. Ctenophora (ctenophores)

4. Platyhelminthes (flatworms)
5. Nemertea or Rhynchocoela (ribbonworms)
6. Acanthocephala (spiny-headed worms)
7. Rotifera (rotifers or wheel animalcules)
8. Gastrotricha (gastrotrichs)
9. Kinorhyncha or Echinoderida (no common name)
10. Preapulida (Priapulids)
11. Nematoda (roundworms)
12. Nematomorpha (horsehair worms)
13. Conodonta (conodonts)
14. Entoprocta or Kamptozoa (entoprocts)
15. Annelida (annelid worms)
16. Arthropoda (arthropods – insects, crabs, shrimps, etc.)
17. Mollusca (mollusks)
18. Bryzoa or Ectoprocta (bryzoans)
19. Phoronida (phoronid worms)
20. Brachiopoda (brachiopods)
21. Sipunculoidea or Sipuncula (sipunculid worms)
22. Chaetognatha (arrowworms)
23. Echinodermata (echinoderms)
24. Hemichordata (hemichordates)
25. Pogonophora (breadworms)
26. Graptolites (graptolites)
27. Chordata (chordates)

Each of the phyla above listed are subdivided into classes, each class into orders, then families, genera, and species. The last named phyla, for example, the chordata, is divided into classes of fish, amphibia, reptiles, birds, and mammals. Mammals are in turn divided into orders of monotremes (platypus and echidna), marsupials, insectivores (shrews, moles, hedgehogs), flying lemurs, chiroptera (bats), primates (lemurs, monkeys, apes, man), edentates (anteaters), pangolins, aardvark, lagomorphs (rabbits, pika), rodents (mice, rats, squirrels, porcupines, beaver), toothed whales, baleen whales, proboscidea (elephants), sirenia (sea cows), hyracoids, perissodactyls (tapirs, rhinoceros, horses), artiodactyls (pigs, cattle, sheep, goats, antelope), *and 19 orders known only in fossils.*

All of the foregoing animals plus all plants developed in the theory of evolution from a glob of protein-like material which formed a living organism programed to reproduce only its own kind. Each subsequent organism was programed in its DNA to reproduce only its own species. All of those changes from a protozoan to produce at least 2,000,000 different species of various types that are as different in their insides as their external appearances have been proposed in face of the fact that not one of the postulated changes has ever been observed to occur. Not only are changes required, but the different offspring must have been isolated with others of their own kind. No species has ever been observed to reproduce anything but the same species. Modern genetics has provided the reason. Furthermore, anatomical, physiological, and chemical similarities of various plants and animals is more consistent with multiple originations in this new theory of chemical origins than with the theory of evolution that required pliability and changes.

Every experienced experimental scientist has subjected hypotheses to sequences of tests. A hypothesis may remain viable or it may be disproven by any of a series of experiments and observations. The theory of evolution has not passed the tests. It conflicts with crucial scientific observations of offspring of animals only repeating the species and chromosome sizes, shapes, and numbers proving common ancestors of species are impossible. It also conflicts with observations of most species being endemic to a given land mass and observations of unusual characteristics of many plants and animals. Its only evidence, fossil sequences, can be explained by changes in atmosphere and changes in conditions that produced living organisms. It is simply time that the theory of evolution was discarded. It is now clear that scientific evidence requires multiple originations of living organisms whether or not the mechanism was as postulated in the present new theory. The new theory is logical and does satisfactorily explain all relevant facts.

XX. HOMO SAPIENS

An objective analysis of how humans fit into the various theories of origins reveals reasons for confusion. *Homo sapiens* has characteristics that are so different that it is difficult to explain its presence as being of the same character as other animals. No other animal has dense hair distributed in a pattern anything close to that of man. No other animal has hair concentrated on its head, neck, face, axillae (armpits), and around the genitalia. No other animal has hair on its head and face that continues to grow longer. With the exception of mimicry ability of certain birds, no other animal makes the sounds of humans. No other animal produces tears when it cries. No other animal laughs. The mental powers of the human are apparently greater than those of other animals, although we may be underevaluating the brains of some animals due to their lack of ability to accumulate and communicate experience as well as man does. The long period of helplessness of a baby is in marked contrast with young of other animals. Man's ability to smell is grossly less than that of many animals. His hearing ability may be average, certainly not superior. Vision capabilities are difficult to compare, but distance vision is clearly less than that of some birds.

The human has many more than its share of diseases. It is recognized that animals have distempers; but no other animals have anything that compare with the frequent runny noses, cold sores, cankers, sore throats, tonsillitis, ear infections, dental caries, measles, mumps, diptheria, small pox, chicken pox, cholera, tetanus, whooping cough, poliomyelitis, rheumatic fever, and other diseases of humans. Cancer occurs in other animals, but the frequency difference is enormous.

Some animals appear to be engineering marvels designed with functional form and beauty. For examples, squirrels, song birds, house cats, tigers, dogs, zebra, horses, and monkeys are animals that do not appear to be the result of side reactions in a vat of chemicals. On the other hand, some animals are awkward and not beauty-contest types. An ostrich can make one smile, but it would not be mistaken for a Frank Lloyd Wright design. Some animals are too large or too heavy to perform well according to the usual purposes of their parts. Other animals appear to be made of spare parts of other animals without consistency of form for function.

Humans have the appearance of design for good function and beauty. They are the only true primates. Monkeys and apes use their fore limbs for locomotion on the ground and in trees. Humans do not appear to be made of spare parts of other animal productions, and other animals do not have human parts in their constructions.

The differences between human beings and other animals is great. They are great when comparing humans with all animals as a group, and they are great when considering them with other animals on the same continent. Monkeys resemble other animals, and look as if they belong to the group. Humans do not. There is no monkey or ape that appears to have changed a basic characteristic or part of his design like other animals toward the design of the human. It would seem strange if an animal as well designed as a monkey or ape would change to something intermediate toward the human that would probably be of poor design for usual animal needs.

In this theory of multiple originations and creations, humans appear to be organisms that were produced separately from other animals. That would account for their differences from other animals. When comparing various races, there are distinct differences which correlate with geography similar to that of other animals. Natives of Central and South America resemble those of the Asian Subcontinent. That resemblance is closer than the resemblance of natives of the Asian Subcontinent and those of contiguous Europe. The African humans are as distinctive as other animals of that continent. There are similarities in human natives of North America and those of Asia just as with other animals.

Other evidence that races of humans originated on separate land masses is deficient or lack of immunity to bacteria and viruses of foreign places. Smallpox from Europe killed American Indians in great numbers. Prior to Columbus's visits to the Americas, neither the American natives or Europeans were acquainted with syphilis. After his visits, the disease ravaged American natives as well as the Europeans. It suggests that both were infected with each other's germs that caused the original bearers no apparent significant harm.

The genital herpes virus affecting Americans that has recently received a great deal of publicity undoubtedly came from the Orient. At the end of the Korean War, in 1952, while serving as a Navy medical officer on a troopship, I was the first to recognize that what was commonly being diagnosed as chancroid was actually a herpes-like disease. It was misdiagnosed by others as chancroid because the lesions were much worse than cold sores and common cankers of the mouth. With multiple recurrences and increasing immunity, the character of the lesions changes to a more benign appearance and the diagnosis of herpes is more easily made. The virus apparently caused no problems to the Japanese and Koreans, but many American sailors and soldiers had little natural immunity. There were other diseases that followed a similar immunity pattern.

It may be noted that although the etiology or cause of most cancers is unknown, that mummies, skeletons of humans buried in Italy in early historical times, skeletons in burial mounds of American Indians built prior to 1492, and fossils of humans have shown no cancer.

It is postulated that some of the varieties or races of *Homo sapiens* originated separately consistent with the theory of multiple originations and creations of other animals. Each of those creations of races of humans occurred under very similar circumstances and without mixing of DNA with that of other animals. Those originations probably occurred at the locations of established petroleum seepages. It is probable that the African black man originated in Africa, that the brown Indo-Egyptians most likely originated in the Middle East and/or the Burma area, and the white race may have originated farther north in Iran, the U.S.S.R. and/or Europe. Any of the races may have had multiple sites of origin. The natives of Central and South America probably originated in that area where large petroleum seepages are located. North American natives probably originated in Southern California where there are numerous petroleum seepages.

Adoptions of humans by established species could have happened by as simple a process as a predator carrying an infant to its lair where instead of becoming a meal it found a nipple.

From the foregoing, in this new theory some or all races of *Homo sapiens* originated separately on various land masses corresponding to the general areas in which they are natives.

XXI. CONCLUSION

A new comprehensive theory is presented which explains interrelations of the substance of Earth, volcanoes, petroleum, the oceans, the atmosphere, and life.

It has been shown that the generally accepted theory that petroleum was produced from formerly living organisms is inconsistent with the facts and is probably wrong. There is no known chemical or biologic process that can convert the matter of formerly living organisms to petroleum.

The organic biologic theory cannot account for the general chemical uniformity of all petroleum wherever it is found and its huge quantities. Ethane, propane, and butane, hydrocarbon gases found with petroleum, cannot be accounted for in the theory. The explanations given for the presences of helium and argon with petroleum based upon radioactivity are unsatisfactory. They do not account for the presence of methane whenever helium is found, and the absence of hydrogen which would be expected from radioactivity acting upon organic matter. The large concentrations of nitrogen and carbon dioxide found with petroleum are inconsistent with the organic theory.

The finding of water with petroleum is inconsistent with the organic theory, and the postulate of independent sources is inconsistent with the concentrations of minerals found in the water. The absence of phosphates in petroleum and oil field water is particularly damaging to the organic theory.

Petroleum is found in geologic strata that have no connections with oil shale, fish, coal, or any other deposits of biologic matter. There are no substances intermediate between petroleum and any of those biologic substances. Deposits of biologic matter on ocean floors do not contain any liquid hydrocarbons.

The findings of porphyrins in petroleum and the optical activity of petroleum which have been heavily relied upon as proof of the biologic origin of petroleum have been accounted for by inorganic, non-biologic processes. They are not evidence of biologic origin of petroleum.

High temperatures are required to produce the chemicals found in petroleum. In the biologic origin of petroleum theory, the temperatures are insufficient for the job. Sufficient heat would be completely inconsistent with the theory and the evidence upon which it relies. Furthermore, there are no residues of organic matter which would have resulted from its having been cooked. The sizes of hydrocarbons and quantities of ring hydrocarbons found in petroleum are also inconsistent with the organic theory.

It has been shown that the organic theory of petroleum production is inconsistent with geologic observations and facts of chemistry. A new theory of inorganic, abiotic origin of petroleum is presented in a comprehensive theory encompassing causes of volcanoes and originations of life.

The new comprehensive theory postulates that when Earth originated gases were released due to the heat of internal pressure and resulting nuclear reactions producing heavy atoms. The gases included water as steam, methane and ammonia corresponding to their presences in atmospheres of Jupiter and Saturn, nitrogen, and carbon dioxide.

As the gases and Earth cooled below the boiling point of water, oceans formed which covered the Earth. The pressure of the water on the crust of the Earth then impeded the escape of gases. Gases accumulated under the crust in sites scattered more or less equidistant from each other about the globe. When enough gas under sufficient pressure had accumulated at each site, it erupted with accompanying magma. Those eruptions formed the nuclei of continents in their approximate present locations, scattered about the globe, with contrasting deepening ocean basins. Gases under the oceans then migrated along the generally upward course to the edges of the oceans. If those gases reached large bodies of sediment capped by impervious compact sediments or rock, they polymerized and became petroleum. If they were not contained, the gases escaped through volcanoes and accompanying magma increased the sizes of the continents. When the Earth was young, and sediments were thin, some gases became converted to petroleum; but with mounting pressures from more gases from the magma, volcanoes erupted that petroleum along with usual volcanic gases. Similarly, the first eruptions of later volcanoes probably contained petroleum. As they become mature volcanoes, usually the only hydrocarbon they erupt is methane in small amounts.

This new theory postulates that when hot, fresh petroleum reached the surface of the earth it reacted chemically with phosphorus, water, and other common chemicals to produce the essential chemicals of living organisms. That included DNA, RNA, and proteins. Petroleum from volcanoes and seepages provided an abundant source of material.

When the earliest living organisms originated, it is postulated that they were abundant. Units of life, DNA and RNA, along with protein, stuck together to their mutual advantage and formed cells. Single cells formed in large numbers included germ cells capable of producing all complex forms of life.

Movements of the crust of the Earth and resulting new juvenile volcanoes and new seepages of petroleum of succeeding geologic periods provided periodic originations of new living organisms, plants and animals.

The sequence of fossils until the modern Cenozoic Era was probably determined by the atmosphere. The only organisms that were able to live at first were anerobic bacteria. The atmosphere was apparently poisonous to land plants for three geologic periods after animals were abundant in oceans. Plants provided free oxygen to the water which allowed shelled animals with low oxygen requirements to thrive. Oxygen was slowly freed from water in the upper atmosphere by the action of ultraviolet light on water. That oxygen oxidized the methane to carbon dioxide, and possibly other substances poisonous to plants were removed from the atmosphere.

When those poisonous substances were removed, plants could live on land and forests grew. By photosynthesis plants provided free oxygen to the atmosphere. Iron on the surface of the Earth has continually combined with the oxygen, increasing the problem of reaching oxygen concentrations adequate for land animals. The land animals that lived were in a sequence determined by their oxygen requirements. Insects, amphibia, and reptiles preceded mammals and finally birds because of their respectively increasing oxygen requirements.

Gradual cooling of the Earth punctuated by colder episodes has probably been responsible for some extinctions of both aquatic and terrestrial flora and fauna. That has been pronounced in the Pleistocene epoch in the modern Cenozoic Era during which the Northern Hemisphere has experienced four extensive glaciations in a span of two and a half million years.

In this theory of multiple originations all species are original. The concept that life originated as a result of combinations of molecules in the atmosphere has been proven wrong by the inability of chemists to produce the necessary chemicals in the postulated conditions. Chromosome findings have revealed the reason why species give rise to only the same species. Chromosome numbers, shapes, and sizes have proven that modern species of the same genera could not have ancestors in common. It has been proven by their chromosomes that none of the lemurs, shrews, monkeys, and apes could be ancestors of humans or that they could be related through common ancestors. The theory of evolution has thereby been proven wrong by scientific criteria. Furthermore, the theory of evolution has failed to account for observations of living plants and animals.

The new theory is simple, direct, and logical. With heat from the interior of the Earth, parts of its substance have volatilized. The release of those gases has been impeded by the crust of the Earth, and particularly where it is overlaid by oceans. The gases have been converted to petroleum where they have reached sediments in which they have been contained. Where they have not been contained, they have escaped through volcanoes. Juvenile volcanoes have erupted mixtures of petroleum and usual volcanic gases. Petroleum has also reached the surface through common seepages and so-called "mud volcanoes." That petroleum has provided material and heat for chemical reactions that have produced the chemicals of life, including DNA, RNA, and protein.

Units of life including DNA and RNA have joined together with protein and other chemicals and have formed single cells including germ cells in variety and in large numbers. In their natural growth, those germ cells have produced all of the species and some subspecies or races.

Extinctions of many plants and animals have occurred probably due mainly to changes in the environment, particularly the temperature, and not primarily due to an evolutionary status.

Periodic movements of the crust of the Earth have caused new seepages of petroleum to the surface and new juvenile volcanoes which have erupted petroleum. The periodic new volancoes and new seepages of petroleum have been recurring phenomena producing the same type of conditions that first gave rise to life on primordial Earth. They have provided periodic originations of new and different plants and animals due to slight changes in conditions and environment of the originations.

The primitive character of the flora and fauna of the Hawaiian Islands and the Galapagos Islands correlates with their geologic history and volcanic nature being similar to primordial continents. The nature of the animals of Australia correlates with the geologic history of that continent which has missed the major tectonic episodes and associated new volcanoes and petroleum seepages of geologic periods in which modern placental mammals originated.

Petroleum seepages have probably provided stability in temperature and nutrients in pools required for gestations of placental mammals, and especially humans. Adoptions have most likely played an important role.

The theory of evolution, which this new theory should replace, has no basis in scientific observation or proven scientific principles. There is absolutely no evidence that animals have become different species than their ancestors. Fossil sequence is explained on the rational basis of oxygen requirements of various animals and changing oxygen concentrations in waters and in the atmosphere. This new theory is based upon sound, proven, scientific observations and principles, common observations, and sound reasoning.

SELECTED BIBLIOGRAPHY

Bell, H.S., Oil Shales and Shale Oils, D. Van Nostrand Co., New York, 1984.

Berger, Andrew J.: Hawaiian Birdlife, The University Press of Hawaii, Honolulu, 1972.

Bernal, J.D.: The Origin of Life, The World Publishing Company, Cleveland, 1967.

Brooks, Benjamin T., Kurtz, Stewart S., Jr., Boord, Cecil E., Schmerling, Louis: The Chemistry of Petroleum Hydrocarbons, Reinhold Publishing Corp., New York, 1954.

Bullard, Fred M.: Volcanoes of the Earth, University of Texas Press, Austin & London, 1976.

Caras, Roger: Venomous Animals of the World, Prentice-Hall, Inc., Englewood Hills, N.J., 1974.

Caras, Roger A.: Dangerous to Man, Rinehart and Winston, New York, 1964.

Campagnat, Alfred: *Protein from Petroleum,* Scientific American, Vol. 213, no. 4, pp. 13-17, Oct. 1965.

Darwin, Charles: Origin of Species and the Descent of Man, The Modern Library, New York.

Degener, Ott: Plants of Hawaii National Park: Illustrations of Plants and Customs of the South Seas, Edwards Brothers, Inc., Ann Arbor, 1945

Dott, Robert H., Sr. and Reynolds, M.J.: Sourcebook for Petroleum Geology, Am. Ass. of Petroleum Geologists, Tulsa, Okla., 1969.

Dunbar, Carl O.: Historical Geology, John Wiley & Sons, Inc., New York, 1949.

Dunning, H.N., Moore, J.W., Denekas, M.O.: *Interfacial Activities and Porphyrin Contents of Petroleum Extracts,* Ind. Eng. Chem. 45: 1759, 1953.

Eibl-Eibesfeldt, Irenäus: Galapagos: The Noah's Ark of the Pacific, trans. by Brodrick, Alan Houghton, Doubleday & Co., Inc., Garden City, 1961.

Fox, Sidney W. and Harada, Kaoru: *The Thermal Copolymerization of Amino Acids Common to Protein,* J. Amer. Chem. Soc., 82 (14): 3745 (1960).

Gardner, Martin: The Ambidextrous Universe: Left, Right, and the Fall of Parity, New American Library, New York, 1969.

Gardner, Eldon J.: Principles of Genetics, 5th Ed., John Wiley & Sons, Inc., New York, 1975.

Gruse, W.A., Stevens, D.R.: The Chemical Technology of Petroleum, McGraw Hill Book Co., New York, 1942.

Gurwitsch, L. and Moore, H.: The Scientific Principle of Petroleum Technology, D. Van Nostrand Co., New York, 1937.

Hitching, Francis: The Neck of the Giraffe: Where Darwin Went Wrong, Ticknor & Fields, 1982.

Hobson, G.D.: Some Fundamentals of Petroleum Geology, Oxford University Press, London, 1954.

Hobson, G.D. and Tiratsoo, E.N.: Introduction to Petroleum Geology, Scientific Press, Ltd., Beaconsfield, England, 1975.

Holden, Raymond: Famous Fossil Finds: Great Discoveries in Palaeontology, Dodd, Mead & Co., New York, 1966.

Hsu, T.C. and Benischke, Kurt: An Atlas of Mammalian Chromosomes, Springer Verlag, New York, Heidelberg, Berlin, 1974.

Jastrow, R., Rasool, S.I.: The Venus Atmosphere, Gordon and Breach, New York, London, Paris, 1969.

Landes, Kenneth K.: Petroleum Geology 2nd Ed., p. 164, John Wiley & Sons, Inc., New York, 1959.

Levorsen, A.I.: Geology of Petroleum, 2nd Ed., W.H. Freeman and Co., San Francisco and London, 1967.

MacDonald, Gordon A.: Volcanoes, Prentice-Hall, Inc., Englewood Cliffs, N.J., 1972.

March, Jerry: Advanced Organic Chemistry: Reactions, Mechanisms, and Structure, McGraw-Hill Book Co., New York, 1968.

Miller, Stanley L. and Orgel, Leslie E.: The Origins of Life on Earth, Prentice-Hall, Inc., Englewood Cliffs, N.J., 1974.

Morris, Percy A.: A Field Guide to Shells of the Pacific Coast and Hawaii, 2nd Ed., Houghton-Mifflin Co., Boston, 1966.

Morrison, J.D., Mosher, H.S.: Asymmetric Organic Reactions, Prentice-Hall, Englewood Cliffs, N.J., 1971.

Oparin, A. and Fesenkov, V.: Life in the Universe, Twayne Publishers, Inc., New York, 1961.

Pauling, L. and Corey, R.B.: *Atomic Coordinates and Structure Factors for Two Helical Configurations of Polypeptide Chains,* Proc. Natl. Acad. Sci. U.S., 37: 235 (1951).

Purdy, G.A.: Petroleum Prehistoric to Petrochemicals, Copp Clark Publishing Co., Vancouver, Toronto, Montreal, 1958.

Report of a Conference on the Origin of Oil Research Committee of the American Association of Petroleum Geologists, Am. Assoc. of Petroleum Geologists, Tulsa, Okla., 1941.

Rittman, A.: Volcanoes and Their Activity, John Wiley & Sons, New York, 1962.

Russell, W.L.: Principles of Petroleum Geology, 2nd Ed., McGraw-Hill Book Co., New York, 1960.

Sachanen, A.N.: The Chemical Constituents of Petroleum, Reinhold Publishing Corp., New York, 1945.

Simpson, George Gaylord: The Geography of Evolution, Chilton Books, Philadelphia & New York, 1965.

Skinner, D.A.: *Chemical State of Vanadium in Santa Maria Valley Crude Oil,* Ind. Eng. Chem. 44: 1159-1165, 1952.

Tiratsoo, E.N.: Oilfields of the World, Scientific Press, Ltd., Beaconsfield, England, 1973.

Tyler, Donald E.: A New and Simple Theory of Gravity, Pruett Press, Inc., Boulder, Colorado, 1970.

Tyler, Donald E.: *Common Contagious Genito-Urinary Diseases with Reference to the Ito-Reenstierna and Frei Tests,* British Journal of Venereal Diseases, V. 33, no. 4, 228-241, Dec., 1957.

Tyler, Donald E.: *Epithelium of Intestinal Type in the Normal Urachus: A New Theory of Vesical Embryology,* Journal of Urology, 92:505, Nov. 1964.

Tyler, Donald E.: *Stratified Squamous Epithelium in the Vesical Trigone and Urethra,* American Journal of Anatomy, 111:319, Nov. 1962.

Tyler, Donald E.: *Urinalysis and Urine Cytology Correlated with the Menstrual Cycle and Age,* American Journal of Obstetrics and Gynecology, 90:147, Sept. 1964.

Tyler, Donald E.: *The Relationship Between the Pressure Sustained by the Various Cardiac Valves and the Relative Frequency of their Involvement in Rheumatic Fever,* American Heart Journal, 51:415, March 1956.

Tyler, Donald E: *Role of the Hyperimmune Reaction in Chancroid,* U.S. Armed Forces Medical Journal, 7: 663, Sept. 1956

Van Wiebe, W.A.: North American and Middle East Oil Fields, Edwards Brothers, Inc., Ann Arbor, Mich., 1950.

Wald, George: *The Origin of Optical Activity,* Ann. N.Y. Acad. Sci., 69: 352, Aug. 30, 1957.

Walker, James C.G.: Evolution of the Atmosphere, Macmillan Publishing Co., Inc., New York, 1977.

Wendt, Herbert: Before the Deluge, trans. by Winston, Richard and Clara, Doubleday, Inc., 1968.

Wilson, A.T.: *Origin of Petroleum and the Composition of the Lunar Maria,* Nature, Vol. 196, no. 4849, pp. 11-13, 1962.